De Gruyter Graduate

Benvenuto • Metals and Alloys

Also of Interest

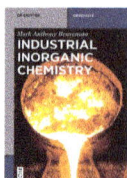

Industrial Inorganic Chemistry
Benvenuto; 2015
ISBN 978-3-11-033032-8, e-ISBN 978-3-11-033033-5

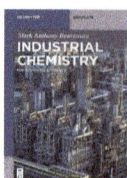

Industrial Chemistry: For Advanced Students
Benvenuto; 2015
ISBN 978-3-11-035169-9, e-ISBN 978-3-11-035170-5

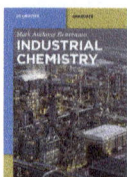

Industrial Chemistry
Benvenuto; 2013
ISBN 978-3-11-029589-4, e-ISBN 978-3-11-029590-0

Corrosion for Engineers
Zander, Dietzel, Atrens; 2016
ISBN 978-3-11-030666-8, e-ISBN 978-3-11-030687-3

Process Technology: An Introduction
De Haan; 2015
ISBN 978-3-11-033671-9, e-ISBN 978-3-11-033672-6

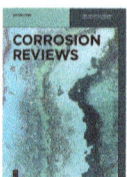

Corrosion Reviews
Latanision, Eliaz (Editors-in-Chief)
ISSN 0334-6005, e-ISSN 2191-0316

Mark A. Benvenuto

Metals and Alloys

Industrial Applications

DE GRUYTER

Author
Prof. Dr. Mark Anthony Benvenuto
Department of Chemistry and Biochemistry
University of Detroit Mercy
4001 W. McNichols Road
Detroit, MI 48221-3038
USA

ISBN 978-3-11-040784-6
e-ISBN (PDF) 978-3-11-044185-7
e-ISBN (EPUB) 978-3-11-043352-4

Library of Congress Cataloging-in-Publication Data
A CIP catalog record for this book has been applied for at the Library of Congress.

Bibliographic information published by the Deutsche Nationalbibliothek
The Deutsche Nationalbibliothek lists this publication in the Deutsche Nationalbibliografie;
detailed bibliographic data are available on the Internet at http://dnb.dnb.de.

© 2016 Walter de Gruyter GmbH, Berlin/Boston
Cover image: Andreas Krone/iStock/Thinkstock
Typesetting: le-tex publishing services GmbH, Leipzig
Printing and binding: CPI books GmbH, Leck
♾ Printed on acid-free paper
Printed in Germany

www.degruyter.com

Preface

The history of many civilizations is often defined in terms of a metal. In many societies, what is called a 'Bronze Age' comes before an 'Iron Age,' and the ability to work and use these metals affects great changes both in the peoples that produce them as well as among neighboring people who traded with the bronze and iron producers. An example from the ancient world is that societies that used iron weapons were routinely able to conquer those still using bronze weaponry. In more modern times, the use of iron farm equipment in the nineteenth century in an expanding United States and Canada allowed huge areas to be turned to agriculture that could not be worked previously with any other type of plow. Indeed, among the many inventions and advances that mark humankind, the production of refined metals ranks alongside the invention of the wheel, the shift from hunting and gathering to farming, and the directing and use of fresh water in terms of overall importance.

This book discusses numerous metals and alloys, both in terms of how they are made, as well as how they are used and applied. Many of the chapters are metals with which most people are very familiar, such as iron, copper, aluminum, gold and silver. Among those five, only aluminum can be said to have a modern history, yet all have influenced history. The search for gold especially has sent people to the ends of the Earth: the Australian Outback, the Yukon, the Congo, and Siberia for example – as well as across what was called in the year 1491 the 'Ocean Sea'.

Other metals are produced on a relatively large scale, but remain relatively obscure to people today, at least in terms of their uses. Examples include titanium, tantalum, niobium, and the lanthanides. But all the metals examined in this book have applications that help enable people today to live a quality of life that humanity has not previously achieved.

The production of so much highly refined metal, and the wealth of uses to which modern metals are now put, has invariably led to a certain amount of pollution and by-product production. Each chapter in this book discusses the current state of recycling for the metal discussed therein. For example, the recycling and re-use of iron is a mature and large-scale industry in itself. Scrap yards the world over deal in buying and selling many grades of used iron and steel. Some countries, such as Bangladesh, have developed industries that cater to the dismantling of large ships so that the iron and other metals in them can be recycled into other products. Yet other metals have yet to have any recycling programs developed for them. Examples include neodymium, used in state-of-the-art cell phones, and americium, used in small amounts in residential and commercial smoke detectors.

This book examines metal production as well as that of alloys in terms of the chemistry involved from ore to refined metal. It is not a chemical engineering text, but it must include some discussion of how processes are adjusted when they are brought to an industrial scale. The aim is to understand how metals are produced, how alloys are made and determined to be beneficial, and to examine the uses of both. It is hoped that this book will be useful to undergraduate and graduate college audiences in chemistry and related sciences.

Beyond this, I would like to thank my friends and colleagues, many of whom I have pestered with questions in order to make this book more complete. Thanks go to Matt Mio, Klaus Friedrich, Kendra Evans, Kate Lanigan, Mary Lou Caspers, Bob Ross, Shula Schlick, Jon Stevens, Prasad Venugopal, Gary Hillebrand, Liz Roberts-Kirchhoff, and Heinz Plaumann for tolerating my questions and for help they might not even know they have rendered. As well, thanks go to Jane Schley and Meghann Murray for their help and support. And of course, thanks go to my wife Marye, and my sons David and Christian, for putting up with me and all the whining that authors tend to do.

Detroit, June 2016

Mark A. Benvenuto

Contents

Preface —— V

1	**Introduction and overview** —— **1**	
1.1	Historic, ancient —— **1**	
1.2	Large-scale use —— **1**	
1.3	Eighteenth and nineteenth century discoveries —— **2**	
1.4	Modern, niche uses —— **2**	
1.5	Modern, major use metals —— **3**	
1.6	Recycling and re-use —— **3**	

2	**Copper** —— **5**	
2.1	Introduction —— **5**	
2.2	Refining and isolation —— **6**	
2.3	Uses —— **8**	
2.3.1	Piping —— **8**	
2.3.2	Wiring and machinery —— **8**	
2.3.3	Coinage —— **8**	
2.4	Bronze —— **10**	
2.5	Brass —— **10**	
2.6	Other alloys —— **11**	
2.7	Possible substitutes —— **11**	
2.8	Recycling —— **12**	

3	**Tin** —— **13**	
3.1	Introduction —— **13**	
3.2	Refining and isolation —— **13**	
3.3	Tin, uses and applications —— **15**	
3.3.1	Solders —— **15**	
3.3.2	Plating —— **15**	
3.3.3	Niobium–tin —— **16**	
3.3.4	Glass production —— **16**	
3.3.5	Tin in polyvinyl chloride —— **16**	
3.4	Possible substitutes —— **17**	
3.5	Recycling —— **17**	

4	**Zinc** —— **19**	
4.1	Introduction —— **19**	
4.2	Refining and isolation —— **19**	
4.3	Uses —— **21**	

4.3.1   Galvanizing —— 21
4.3.2   Nickel silver —— 21
4.3.3   Solders —— 22
4.3.4   Paint pigments —— 22
4.3.5   Sacrificial anodes —— 23
4.3.6   Batteries —— 23
4.4   Possible substitutes —— 23
4.5   Recycling —— 24

5   Pewter —— 25
5.1   Introduction —— 25
5.2   Production —— 25
5.3   History and traditional uses —— 26
5.4   Modern applications —— 26
5.5   Recycling —— 26

6   Gold —— 29
6.1   Introduction —— 29
6.2   Refining and isolation —— 29
6.3   The carat system – 18, 14, 12, 10 Carat —— 33
6.4   Uses —— 33
6.4.1   Jewelry —— 33
6.4.2   Investment coins —— 34
6.4.3   Electronics —— 35
6.5   White gold —— 35
6.6   Possible substitutes —— 36
6.7   Recycling —— 36

7   Silver —— 37
7.1   Introduction —— 37
7.2   Refining and isolation —— 38
7.3   Uses —— 39
7.3.1   Jewelry —— 39
7.3.2   Silverware —— 40
7.3.3   Investment coins —— 40
7.3.4   Photography —— 40
7.3.5   Batteries —— 41
7.3.6   Electronics —— 41
7.3.7   Ethylene oxide —— 41
7.4   Possible substitutes —— 42
7.5   Recycling —— 42

8	Iron and steel —— 43
8.1	Introduction —— 43
8.2	Ore sources —— 45
8.3	Steel production —— 47
8.4	Uses of iron and steel —— 50
8.5	By-product production —— 51
8.6	Recycling —— 52

9	Platinum group metals —— 55
9.1	Introduction —— 55
9.2	Sources, refining and isolation —— 55
9.3	Uses —— 58
9.3.1	Ruthenium —— 58
9.3.2	Osmium —— 59
9.3.3	Rhodium —— 59
9.3.4	Iridium —— 59
9.3.5	Palladium —— 60
9.3.6	Platinum —— 61
9.4	Possible substitutes —— 62
9.5	Recycling —— 62

10	Nickel —— 65
10.1	Introduction —— 65
10.2	Refining and isolation —— 66
10.3	Uses —— 67
10.3.1	Steels —— 67
10.3.2	Superalloys —— 67
10.3.3	Shape memory alloys —— 68
10.3.4	Plating —— 68
10.3.5	Nickels —— 68
10.4	Possible substitutes —— 69
10.5	Recycling —— 70

11	Aluminum —— 71
11.1	Introduction —— 71
11.2	Refining and isolation, the Hall–Heroult process —— 72
11.3	Uses —— 73
11.4	Possible substitutes —— 74
11.5	Recycling —— 75

12   Titanium —— 77
12.1   Introduction —— 77
12.2   Location and sources —— 77
12.3   Refining and isolation —— 79
12.3.1   The Kroll process —— 79
12.3.2   The Hunter process —— 80
12.4   Uses —— 80
12.4.1   High strength alloys —— 80
12.4.2   Pigments, titanium white or titanium dioxide —— 81
12.5   Possible substitutes —— 81
12.6   Recycling —— 82

13   Magnesium —— 83
13.1   Introduction —— 83
13.2   Refining and isolation —— 83
13.3   Uses —— 85
13.3.1   Elemental magnesium —— 85
13.3.2   Magnesium–aluminum alloys —— 86
13.3.3   Castings —— 86
13.3.4   Automotive —— 87
13.3.5   Aerospace —— 87
13.3.6   Electronic —— 87
13.4   Possible substitutes —— 88
13.5   Recycling —— 88

14   Uranium and thorium —— 89
14.1   Introduction —— 89
14.2   Refining and isolation —— 89
14.2.1   Uranium isolation and refining —— 89
14.2.2   Thorium isolation and refining —— 91
14.3   Uses —— 93
14.3.1   Power generation —— 93
14.3.2   Weaponry —— 93
14.3.3   Compounds and uses —— 93
14.4   Possible substitutes —— 94
14.5   Recycling —— 94

15   Americium —— 97
15.1   Introduction —— 97
15.2   Refining and isolation —— 97
15.3   Uses —— 98
15.3.1   Smoke detectors —— 98

15.3.2	Other uses —— **98**	
15.4	Recycling —— **99**	

16 **Mercury —— 101**
16.1	Introduction —— **101**	
16.2	Refining and isolation —— **101**	
16.3	Uses —— **103**	
16.3.1	Chlor-alkali process —— **103**	
16.3.2	Barometers and thermometers —— **104**	
16.3.3	Amalgams and compounds —— **104**	
16.4	Possible substitutes —— **104**	
16.5	Recycling —— **105**	

17 **Lanthanides —— 107**
17.1	Introduction —— **107**	
17.2	Refining and isolation —— **107**	
17.3	Uses —— **112**	
17.3.1	Catalysts —— **113**	
17.3.2	Magnets, $Nd_2Fe_{14}B$, plus $SmCo_5$ and Sm_2Co_{17} —— **113**	
17.3.3	Alloys —— **114**	
17.3.4	Heavy rare earth element uses —— **115**	
17.4	Possible substitutes —— **115**	
17.5	Recycling —— **115**	

18 **Lead —— 117**
18.1	Introduction —— **117**	
18.2	History —— **117**	
18.3	Refining and isolation —— **118**	
18.4	Uses —— **121**	
18.4.1	Batteries —— **121**	
18.4.2	Construction material —— **121**	
18.4.3	Ammunition, hunting and military —— **122**	
18.4.4	Alloys —— **122**	
18.5	Possible substitutes —— **122**	
18.6	Recycling —— **122**	

19 **Tungsten —— 125**
19.1	Introduction —— **125**	
19.2	Refining and isolation —— **129**	
19.3	Uses —— **130**	
19.3.1	Carbide parts —— **130**	
19.3.2	High-strength alloys —— **131**	

19.4   Possible substitutes —— 131
19.5   Recycling —— 131

20   Tantalum and niobium —— 133
20.1   Introduction —— 133
20.2   Refining and isolation —— 135
20.3   Uses —— 136
20.3.1   Major uses —— 136
20.3.2   Wiring and magnets —— 137
20.3.3   Niobium coinage —— 137
20.4   Possible substitutes —— 137
20.5   Recycling —— 138

21   Sodium —— 139
21.1   Introduction —— 139
21.2   Refining and isolation —— 140
21.2.1   Downs cell —— 140
21.2.2   Chlor-alkali process —— 141
21.3   Uses —— 141
21.3.1   Sodium borohydride —— 142
21.3.2   Sodium azide, NaN_3 —— 142
21.3.3   Triphenylphosphine, $P(C_6H_5)_3$ —— 143
21.3.4   Reactor moderator material —— 143
21.3.5   Alloying agent —— 143
21.3.6   Sodium vapor lamps —— 143
21.4   Recycling —— 144

22   Lithium —— 145
22.1   Introduction —— 145
22.2   Refining and isolation —— 145
22.3   Uses —— 146
22.3.1   Lightweight alloys —— 147
22.3.2   Batteries —— 147
22.4   Possible substitutes —— 148
22.5   Recycling —— 148

Index —— 151

1 Introduction and overview

1.1 Historic, ancient

Metals have shaped human societies and human history since before recorded history. Many societies pass through what is called a Bronze Age before entering an Iron Age, a phenomenon identified by the melting points of a metal alloy and an elemental metal, respectively. The fact that the melting point of iron is significantly higher than the melting points of bronze alloys means that iron ages occur after bronze ages, as the ability to make higher temperature fires was achieved in a particular civilization.

Several ancient societies have made extensive use of different elemental metals or alloys, and are still noted for them today. For example, Han Dynasty bronze urns and other objects reached such an apex of the metallurgist's art that these ancient bronzes continue to command high prices in the auction houses of the world's art market today. In ancient Egypt, some of the earliest iron that was worked came from meteorites, and was made into knives that were rather romantically called 'daggers from heaven' by their users. As well, the rise of Rome from a republic to an empire corresponded with the expanded smelting of iron and its widespread use in weaponry and tools within its territories, so much so that pollution from it has been recovered from the Greenland ice cap. Other civilizations and their advances are also marked by the increased use of bronze and iron.

Some civilizations have used metals extensively, but have not specifically gone through bronze and iron ages. Several of the pre-Columbian societies of Central and South America, for example, were skilled at the use of gold, but never appear to have worked iron. Gold objects from the Aztec, the Inca, the Moche, the Chimu, and peoples who inhabited the Amazon Basin at the time of European first contact were noted by European chroniclers in their journals and memoirs [1].

1.2 Large-scale use

Prior to the Industrial Revolution, the production of metals was conducted on a relatively small scale, utilizing a great deal of human labor at each step of the production and refining process. While such processes rose to what can be considered larger than cottage industries, metal working on an industrial scale only became feasible when machinery and heating apparatuses were developed that could extend the range of human abilities beyond what could be done by a single individual with an anvil, forge, and wood-, dung-, or even coal-fed fire.

Britain was an early producer of iron on a large scale as the Industrial Revolution started, in large part because the country also had abundant supplies of coal, which was needed to produce hot enough furnaces to smelt and reduce the iron. In similar manner, the Falun mine in Sweden has produced copper for over a millennium, and its production peaked during the onset of the Industrial Revolution.

For almost all productions of metals from their ores, people have had to mine the raw metal or the ore. This has led to the creation of several national or international mining organizations that advocate for the industry and its products [2–7].

1.3 Eighteenth and nineteenth century discoveries

An examination of the periodic table indicates that many of the metal elements were only discovered in the past 250 years. Figure 1.1 shows a periodic table with the generally accepted year of discovery and isolation of each element listed in it. The letter 'A' indicates an element that was known from ancient times [8].

It can be seen that several metals and a few nonmetals were indeed known to the ancients, but the largest number of elements that we now consider industrially important were discovered between 1700 and 1900.

The discovery of numerous elemental metals in the nineteenth century did not always coincide with an immediate need or use for them, however. For example, neodymium was first discovered in 1885. However, the neodymium–iron–boron magnets ($Nd_2Fe_{14}B$) that have become ubiquitous in numerous applications where small, strong, permanent magnets are required, have only moved into large-scale production in the past two decades, and indeed were only first produced by General Motors in 1982. Likewise, scandium was first discovered in 1879, and while today there are several niche uses for it, sources estimate that no more than 10–15 tons are used annually [9].

1.4 Modern, niche uses

Several metals and alloys exist that are important in some way, but that are still produced on a relatively small scale. For example, beryllium is only produced by three countries, and in relatively small quantities (tons, instead of tens of tons or thousands of tons), but has become extremely useful in what are referred to as X-ray windows because of its transparency to X-rays. Likewise, metals such as the just-mentioned scandium, as well as technetium and rhenium are all produced on a small scale, but all find uses in some niche in one or more specific industries. More recently, tiny amounts of americium have found use in most residences in the developed world.

1																	2
H 1766																	He 1895
3 Li 1821	4 Be 1828											5 B 1808	6 C A	7 N 1772	8 O 1771	9 F 1886	10 Ne 1898
11 Na 1807	12 Mg 1808											13 Al 1825	14 Si 1824	15 P 1669	16 S A	17 Cl 1808	18 Ar 1894
19 K 1807	20 Ca 1808	21 Sc 1879	22 Ti 1791–1910	23 V 1801	24 Cr 1798	25 Mn 1774	26 Fe A	27 Co 1732	28 Ni 1751	29 Cu A	30 Zn A	31 Ga 1875	32 Ge 1886	33 As 1250	34 Se 1817	35 Br 1825	36 Kr 1898
37 Rb 1861	38 Sr 1808	39 Y 1840	40 Zr 1798	41 Nb 1864	42 Mo 1781	43 Tc 1937	44 Ru 1844	45 Rh 1804	46 Pd 1803	47 Ag A	48 Cd 1817	49 In 1867	50 Sn A	51 Sb A	52 Te 1782	53 I 1811	54 Xe 1898
55 Cs 1882	56 Ba 1808	72 Hf 1922	73 Ta 1802	74 W 1783	75 Re 1925	76 Os 1803	77 Ir 1803	78 Pt 1735	79 Au A	80 Hg A	81 Tl 1862	82 Pb A	83 Bi 1753	84 Po 1902	85 At 1940	86 Rn 1910	
[87] FR 1939	[88] Ra 1902																

57 La 1839	58 Ce 1803	59 Pr 1885	60 Nd 1885	61 Pm 1945	62 Sm 1879	63 Eu 1901	64 Gd 1886	65 Tb 1842	66 Dy 1886	67 Ho 1878	68 Er 1842	69 Tm 1879	70 Yb 1907	71 Lu 1906
[89] Ac 1899	[90] Th 1829	[91] Pa 1913	[92] U 1789	[93] Np 1940	[94] Pu 1941	[95] Am 1944	[96] Cm 1944	[97] Bk 1949	[98] Cf 1950	[99] Es 1952	[100] Fm 1952	[101] Md 1955	[102] No 1958	[103] Lr 1961

Fig. 1.1: Periodic table with year of discovery of the elements.

1.5 Modern, major use metals

Iron, copper, aluminum, silver, gold, lead, tin, zinc, nickel and tungsten are all metals that find enormous uses in our modern world. Some of these, such as iron, have been used since ancient times, but now fill much larger roles, such as the construction of bridges and the reinforcing structures of high-rise buildings. Likewise, copper has been used for millennia, but only in the past two centuries has it been utilized in the tens of thousands of miles of wiring that delivers electricity to homes, businesses, and other concerns of the modern world.

Unlike iron and copper, aluminum has gone from discovery in 1825 to use in thousands of applications today. In this case, the production of inexpensive electricity was the key to making large-scale refining of aluminum an economically feasible idea. Similarly, tungsten has a relatively short history, but has found widespread use in a variety of alloys today.

1.6 Recycling and re-use

In almost every case, it is economically beneficial to recycle refined metals when one compares the cost to that of refining new metals from ores. Indeed, some metal recycling has gone on for a century, with metal drives being a part of national economies in both world wars. Today, developed nations throughout the world often mandate some form of metal recycling – often of various food and beverage cans – although different states and provinces in different countries may not have the same requirements and standards. Also, some metals are used in such small amounts in end-user products that there are not yet recycling programs for them. The relatively new recycling of cell phones may be an attempt to remedy that situation when it comes to neodymium, for example.

This book will discuss recycling when possible, from metals that are produced on enormous scales, to metals that are produced in small amounts, but that serve some vital niche use.

Bibliography

[1] The Discovery Of The Amazon: According To The Account Of Friar Gaspar De Carvajal And Other Documents (American Geographical Society), 2007, ISBN: 978-1432557195.
[2] International Council on Mining and Metals. Website. (Accessed 13 April 2015, as: http://www.icmm.com/).
[3] Mining Association of Canada. Website. (Accessed 15 April 2015, as: http://mining.ca/).
[4] Australian Mines and Metals Association. Website. (Accessed 15 April 2015, as: http://www.amma.org.au/).
[5] European Association of Mining Industries, Metal Ores and Industrial Minerals. Website. (Accessed 15 April 2015, as: http://www.euromines.org/).
[6] National Mining Association. Website. (Accessed 15 April 2015, as: http://www.nma.org/).
[7] 40 Common Minerals and Their Uses. Website. (Accessed 15 April 2015, as: http://www.nma.org/index.php/minerals-publications/40-common-minerals-and-their-uses#).
[8] International Women In Mining Community. Website. (Accessed 16 April 2015, as: http://womeninmining.net/library/associations-by-mineral-metal/).
[9] United States Geological Survey. Mineral Commodity Summaries, 2015. Website. (Accessed 24 June 2015, as http://minerals.usgs.gov/minerals/pubs/mcs/2015/mcs2015.pdf).

2 Copper

2.1 Introduction

Copper and its alloys have an ancient history in many cultures. The island of Cyprus gives its name to the metal because so much was used in ancient Greece and Middle East, but as mentioned in Chapter 1, the Han Dynasty of ancient China also used enough copper – in its alloy form, bronze – that Han bronzes command high prices in the art world even today. Copper and its two major alloys, brass and bronze, have been instrumental in advancing civilizations, because tools and weapons made from these metals enable people to farm more efficiently than if they used wooden plows and tools, and because bronze weapons have proven to be more effective than bone, wood and stone weapons.

There are numerous companies throughout the world that produce copper today. The top ten are listed in Table 2.1, although many more exist. Additionally, a large num-

Table 2.1: Top ten worldwide producers of copper [2–11].

No.	Name	Location	Amt. (1000s metric tons)	Other products
1	Codelco	Chile	1,757	
2	Freeport-McMoRan Copper & Gold Inc.	USA	1,441	Gold
3	BHP Billiton Ltd.	Australia	1,135	Aluminum, iron, manganese, nickel, silver, titanium, uranium
4	Xstrata Plc	Switzerland	907	Coal, ferrochrome, nickel, zinc
5	Rio Tinto Group	UK/Australia	701	Gold
6	Anglo American Plc	UK	645	Iron, manganese, platinum, diamonds, nickel, phosphates
7	Grupo Mexico	Mexico	598	Gold, silver, molybdenum, zinc, lead
8	Glencore International AG	Switzerland	542	Gold
9	Southern Copper Corp.	USA	487	Silver, zinc
10	KGHM Polska Miedz	Poland	426	Silver

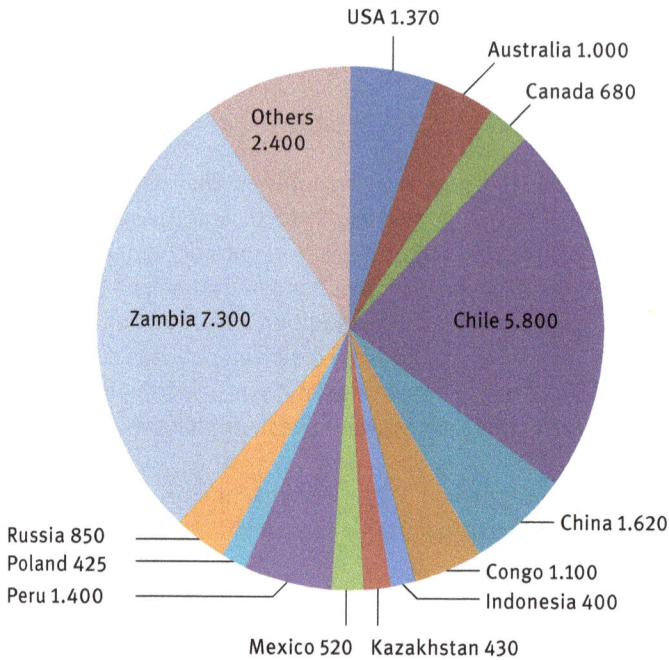

Fig. 2.1: Copper production, in thousands of metric tons.

ber of companies produce copper as one of their secondary metals (such companies that usually mine gold or another more valuable metal as their first product).

Copper continues to be mined in many countries throughout the world, because it finds use in so many different applications. Copper wiring and piping come quickly to mind when people think of uses for the metal, but there are many others as well. Figure 2.1 shows a recent breakdown of where copper is mined.

It can be seen that Chile and Zambia currently dominate the world's production of copper, but that it is widely spread about the world, on all six inhabited continents [1]. Within the United States, copper production takes place primarily in the western states of Arizona, Montana, Nevada, New Mexico, and Utah.

2.2 Refining and isolation

Copper can be found near the Earth's surface, and in such locations it is called placer copper. However, most copper is mined, often as an ore, or as a combination of different copper-containing ores. Table 2.2 is a noncomprehensive list of copper ores.

The term 'porphyry' is also used when discussing copper ores, but this material is a mix of two or more of the copper minerals that have been listed. Still, like the others, it can be processed into the reduced metal.

Table 2.2: Copper ores.

Name	Formula unit	% Copper	Locations (examples)
Bornite	$2Cu_2 \cdot FeS \cdot CuS$	63.3	Kazakhstan, Congo
Chalcocite	Cu_2S	79.8	Australia
Chalcopyrite	$FeCuS_2$	34.5	USA, Canada, Norway
Covellite	CuS	66.5	USA, Austria, New Zealand
Cuprite	Cu_2O	88.8	Russia, Italy, USA
Malachite	$CuCO_3 \cdot Cu(OH)_2$	57.3	Congo, Russia, USA, Zambia
Tennantite	$Cu_{12}As_4S_{13}$	51.6	Congo, Namibia, USA

The isolation of copper from its ores may differ in its steps from one ore batch to another, but general steps include the following:

1. Mining and crushing. Ores or copper deposits are mined, then ground and crushed into small pieces that are easily workable, usually 1–2 cm in size.
2. Grinding. This step reduces and homogenizes particle size until it is a powder. This maximizes surface area of the starting material, and enhances the efficiency of the upcoming steps.
3. Concentration and beneficiation. Since ore concentrations can be low, water is used to extract materials that contain no copper, thus concentrating the usable ore.
4. Smelting. Sulfide ores must be smelted to drive off the sulfur. Oxide ores can be concentrated through solvent extraction. This step may need to be repeated, but can produce copper at 99 % purity.

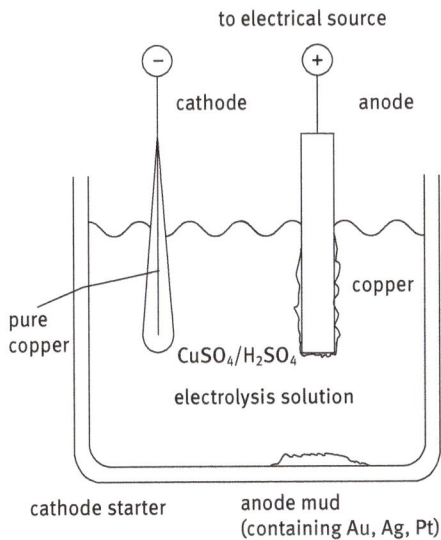

Fig. 2.2: Production of high-purity copper.

5. Leaching. Copper oxide ores can be made into sulfate solutions using an acid solution to enhance the solubility of the ore.
6. Refinement. Copper that has been smelted can be cast into ingots that can be used as anodes. Copper solutions can be electroplated to deposit solid copper.
7. Electrowinning, or electrolytic processing. Copper anodes are immersed in a process tank along with extremely pure copper cathodes, and an electric current is passed through the bath. This allows the deposition of extremely pure copper at the cathodes, and the formation of what gets called anode mud. This residue may contain gold, silver, or platinum group metals, and is recovered because of its economic value. The resulting pure copper cathodes can become as large as 300 pounds. Figure 2.2 shows a diagram of the electrolytic refining of copper.

2.3 Uses

The uses of bronze and brass are discussed in Sections 2.4 and 2.5. The uses of copper, while many, can be categorized as follows.

2.3.1 Piping

Copper piping is used in many locations and applications, from indoor residential plumbing to large industrial concerns. The Copper Development Association maintains a 'Copper Tube Handbook', which is available as a downloadable document, and which details copper tubing not only by sizes, but by pressure ratings and by applications (such as heating systems, snow-melting systems, and agricultural sprinkler systems) [12].

2.3.2 Wiring and machinery

Copper is used in enough applications that copper wire is now standardized into what are called American Wire Gauges (or AWG). Lower numbers equate to larger wires with lower electrical resistance. Thus, 12AWG is a larger wire than 14AWG. Homes in the United States typically have wires from 10AWG to 16AWG. Additionally, copper wire is used in producing electromagnetic motors.

2.3.3 Coinage

Most people think of copper use in coins in terms of the smallest denominations, such as one cent or one penny coins in North America, Europe, and the Pacific. However, throughout history copper has been used in gold and silver coins as well, because the resulting alloy is harder than the metal alone. For example, the United States issued gold coins from the 1790s until 1933, and used a 90 % gold, 10 % copper alloy starting in 1838. Additionally, Great Britain issued gold sovereigns from different mints throughout the world, with the composition 91.67 % gold and 8.33 % copper.

Silver coins have often been alloyed with copper, again because the resulting alloy is harder and more durable than elemental silver. An older term, 'coin silver', means an alloy of 90 % silver and 10 % copper. This was lower than the 92.5 % silver and 7.5 % copper alloy called 'sterling silver' on purpose, so there would be no incentive for counterfeiters to melt silverware to produce fake coins.

Several modern coins contain copper in alloy form, and thus do not appear to be copper in color. United States five-cent pieces – nickels – are 75 % copper and only 25 % nickel. The €0.10, €0.20, and €0.50 Euro coins are made of what is called 'Nordic gold', a brass alloy containing 89 % copper, as well as 5 % zinc, 5 % aluminum, and 1 % tin. Likewise, what are called the Golden Dollar coins of the United States are made from brass, although in this case an alloy containing manganese that produces the same electric signature in a vending machine as previous dollar coins. Figure 2.3 shows several of these modern coins.

Fig. 2.3: Modern coin examples utilizing copper.

2.4 Bronze

Bronze, a series of alloys of copper and tin, tends to find applications in sculpture, and traditionally in bell making. Until the nineteenth century, artillery pieces tended to be made of bronze, both for field use by armies, and on-ship use by navies. These large guns were poured as a single piece, then bored down the center, and were loaded through the firing end, also called the muzzle end, by a trained crew. They continued to be used by military forces until the advent of steel guns, and what is called breech loading (loading from the rear of the gun).

Bells continue to be made of bronze simply because the metal produces excellent ringing tones when properly shaped.

Additionally, bronze finds widespread use in grave markers and funerary urns because it weathers slowly and to an attractive patina over long periods of time [13].

Bronze parts for machinery tend to find uses based on performance. If a bronze piece performs better in a specific application than any other material, it will be used. This is generally both a matter of superior performance and possibly cost savings.

2.5 Brass

Brass is a series of alloys of copper and zinc, often with another element or elements included. Throughout history, the terms brass and bronze have been used

Table 2.3: Brass compositions (empty boxes mean an element is absent from a composition).

Brass Name	Cu	Zn	Sn	Fe	Mn	Ni	Al	Pb	Application(s)
Admiralty	69	30	1						Numerous
Aich's alloy	60.66	36.58	1.02	1.74					Seawater environments
Alpha brass	65	< 35							Imitation gilding
Beta brass	50–55	45–50							Castable
Cartridge brass	70	30							Ammunition cases
Gilding metal	95	5							Ammunition
High brass	65	35							Rivets, screws
Low brass	80	20							Metal adapters
Manganese brass	70	29			1.3				US golden dollar coins
Muntz metal	60	40		< 1					Seawater environments
Nickel-brass	70	24.5				5.5			UK £1 coins
Nordic gold	89	5	1				5		€0.10, €0.20, €0.50 coins
Prince's metal	75	25							Jewelry
Red brass	85	5	5					5	C23000, cast parts
Rivet brass	63	37							
Tombac	85	15							Jewelry
Yellow brass	67	33							Household adornments

interchangeably, in part because before the eighteenth century it is not clear whether craftsmen always knew what elements they were including when producing a specific alloy.

Currently, industry utilizes a wide variety of brass compositions. There is no over-arching theory of what brass composition to use for a specific application. Rather, specific alloys have simply been found to be effective in one application or another. Table 2.3 shows a list of several of the common ones, and their usual application(s).

2.6 Other alloys

Pewter is an alloy that can contain copper, but is treated in Chapter 5, since copper is not its major component.

Copper–nickel alloys find many uses in different industries. Their use in saltwater environments is noteworthy. The Copper Development Association states: "Copper-nickel (also known as cupronickel) alloys are widely used for marine applications due to their excellent resistance to seawater corrosion, low macrofouling rates, and good fabricability [12]." The Association further comments that 90–10 copper–nickel alloy is one of the most commonly used, because of its resistance to corrosion. This numeric designation means 90 % copper and 10 % nickel. Table 2.3 shows a copper-nickel brass alloy that is used in coinage, again because of its resistance to wear.

2.7 Possible substitutes

Because copper has been used in so many applications, some now for hundreds of years, and because the price of copper has risen on world markets in the past fifty years, the United States Geological Survey (USGS) Mineral Commodity Summaries specifically lists possible substitutes. It states: "Aluminum substitutes for copper in power cables, electrical equipment, automobile radiators, and cooling and refrigera-tion tube; titanium and steel are used in heat exchangers; optical fiber substitutes for copper in telecommunication applications; and plastics substitute for copper in water pipe, drain pipe, and plumbing fixtures [1]."

These substitutes are not immediate replacements for copper in all applications. Usually, they compete with copper based on market prices for various metals. The one exception to this is fiber optical cabling, which has replaced copper in several applications, such as in internet connections.

2.8 Recycling

The recycling of copper, as well as of different grades of brass and bronze, are mature industries throughout most of the world. The price of copper is noted, usually by the pound, in major newspapers, and on the world commodity markets. In most cases, this dictates the price when metals are brought for recycling. Recycling of copper, brass and bronze almost always involves an economic incentive, as it is cheaper to recycle these metals then to produce new metal from ores.

Bibliography

[1]   USGS Mineral Commodity Summaries. (Downloaded as: http://minerals.usgs.gov/minerals/pubs/mcs/2015/mcs2015.pdf).
[2]   Codelco. Website. (Accessed 20 April 2015, as: www.codelco.com).
[3]   Freeport-McMoRan Copper & Gold Inc. Website. (Accessed 20 April 2015, as: www.fcx.com).
[4]   BHP Billiton. Website. (Accessed 20 April 2015, as: www.bhpbilliton.com).
[5]   Xstrata plc. Website. (Accessed 20 April 2015, as: www.xtratacopper.com).
[6]   Rio Tinto Group. Website. (Accessed 20 April 2015, as: www.riotinto.com).
[7]   Anglo American Plc. Website. (Accessed 20 April 2015, as: www.angloamerican.com).
[8]   Grupo Mexico. Website. (Accessed 20 April 2015, as: www.gmexico.com).
[9]   Glencore International SA. Website. (Accessed 20 April 2015, as: www.glencore.com).
[10] Southern Copper Corp. Website. (Accessed 20 April 2015, as: www.southerncoppercorp.com).
[11] KGHM Polska Miedz. Website. (Accessed 20 April 2015, as: www.kghm.pl).
[12] Copper Development Association, Inc. Website. (Accessed 22 July 2015, as: http://www.copper.org/).
[13] Bronze.net. Website. (Accessed 22 July 2015, as: http://bronze.net).

3 Tin

3.1 Introduction

Tin is mined and refined in a wide number of countries, and has been for thousands of years. It is one of the metals known and worked in antiquity. Table 3.1 lists the top ten producers of tin as of 2015, as well as their location. Note that even though one national location is listed for each company, most of these are international corporations, with operations in several areas, possibly even on several continents. Currently, tin is not mined in the United States, although there are reserves in Alaska [1].

3.2 Refining and isolation

In many parts of the world, tin is co-located with other metals, which is why the companies listed in Table 3.1 also appear in the top ten for several other elements. The USGS Mineral Commodity Summaries does track the production and use of tin, since there are several uses that may have some importance on a national level. Figure 3.1 shows tin production worldwide, in metric tons [1].

It is obvious from Figure 3.1 that Indonesia and China are currently the world's major producers of tin. In both cases, a large tin-containing formation sometimes called the East Asian Tin Belt runs roughly from southern China southward to Indonesia. Thus, the nations of China, Thailand, Burma, Malaysia, and Vietnam all extract tin ore from the same large source. In China, tin mining has its headquarters in Kunming, in southern China – although Chinese companies can be state owned and run, and thus include mines in all parts of the country.

Table 3.1: Top ten worldwide producers of tin [2–11].

No.	Name	Location	Amount (tons)	Other products
1	Yunnan Tin Group	China	70,373	PGM, copper, lead, zinc, indium, bismuth
2	Malaysia Smelting Corp.	Malaysia	32,668	
3	Minsur	Peru	24,397	Lead, antimony, copper
4	PT Timah	Indonesia	23,718	Tin chemicals
5	Thaisarco	Thailand	22,986	Solders
6	Yunnan Chengfeng	China	18,300	Antimony, bismuth, copper, gold, lead, silver
7	Guangxi China Tin Group	China	11,870	Tin alloys
8	Empresa Metalurgica Vinto	Bolivia	11,253	
9	Metallo Chimique	Belgium	10,344	Copper, lead, nickel, zinc
10	Gejiu Zi Li*	China	6,000	Gold, silver

* Partially acquired by Yunnan Tin.

Thailand 200　Vietnam 5.400
Rwanda 2.000　Others 100
Russia 600　　Australia 6.100
Peru 23.700　　　Bolivia 18.000
Nigeria 500
Malaysia 3.500　　　Brazil 12.000
Laos 800　　　　Burma 11.000

Indonesia
84.000

China 125.000

Congo 3.000

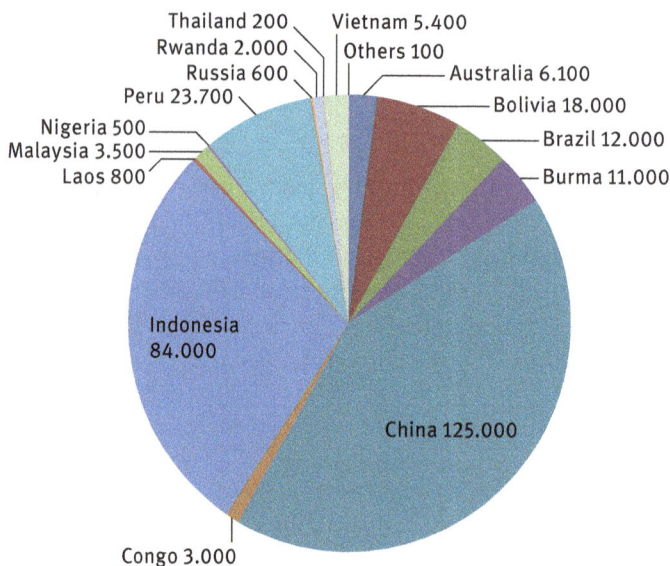

Fig. 3.1: Worldwide tin production, in metric tons.

Likewise, in South America, the famous Potosi mines of Bolivia, which produced enough silver in the sixteenth century that they changed the economy of Europe, began mining tin as a major product in the later nineteenth century, when the price of silver dropped. Both tin and silver can still be mined from these sources.

The major ore of tin, SnO_2, cassiterite (sometimes still called tinstone), must often be concentrated from surrounding rock, other minerals, and what is called overburden – material that does not contain the desired ore.

After concentration, the ore is roasted at elevated temperatures, generally 600 °C, to drive off sulfur-containing materials. This is usually sufficient to produce SnO_2, although depending on each batch, iron oxides can be present as well. The resulting material is then usually leached with water, to dissolve any soluble materials, isolating a purer SnO_2 as an insoluble precipitate.

The chemistry for tin reduction from its ores from this point is straightforward, even though there are variations in ores that are economically feasible to mine. This is one reason that tin has been used in so many cultures for so long – it's extraction from various ores is not particularly complicated. Scheme 3.1 shows the simplified, idealized reaction.

$$SnO_2 + C \rightarrow CO_{2(g)} + Sn_{(l)}$$

Scheme 3.1: Tin refining and reduction.

While this reaction appears simple, it does not include the addition of limestone, which acts as a flux and aids in the formation of slag that is less dense than the reduced metal, and thus easily separated from it. Also, the reaction is routinely run at 1,300 °C–1,400 °C for up to 15 hours, and must be run in some form of refractory-lined furnace, much like the blast furnaces used for iron production.

3.3 Tin, uses and applications

Tin finds numerous uses, as an elemental metal, an alloy, or in organometallic compounds. The USGS Mineral Commodity Summaries notes: "The major uses for tin were cans and containers, 23 %; construction, 18 %; transportation, 17 %; electrical, 12 %; and other, 30 % [1]." Several of these applications are discussed below.

3.3.1 Solders

For more than a century, tin has been used with several other metals, notably lead, to form solders for use in connecting two points in some metal item. This is its principle electrically-related use. The reason for this widespread use is that the mixtures of the two metals melt at lower temperatures than either of the elemental metals alone. Tin has a melting point of 232 °C and lead has a melting point of 328 °C. A mixture that is 55 % tin and 45 % lead by mass however, what is called the eutectic point for the alloy, has a melting point of 176 °C. And interestingly, if there are impurities in either of the metals to as small a total of 1 %, the eutectic point depresses to below 100 °C, meaning that some impure lead–tin solders will melt in boiling water.

There are numerous lead–tin solders that are commercially available, but in the past decade, attempts have been made to find lead-free substitutes. In general, lead–tin solders and others with relatively low melting points are called soft solders, because they melt below 350 °C. Those with higher melting points, such as silver–tin solders, are referred to as hard solders.

3.3.2 Plating

Tin finds use as a plating metal for several other common metals, including steel, zinc, and lead, because it enhances resistance to corrosion of the object to which it is plated. This is a major use of tin in terms of the production of end-use items. Often, tin is plated onto metal food containers, because it is nontoxic and its direct contact with food causes no ill effects.

3.3.3 Niobium–tin

In 1954, an alloy composed of 75 % niobium and 25 % tin (Nb_3Sn) was found to be superconducting. Unfortunately, the expense and difficulty in drawing the material into a wire for production of magnetic coils has limited its application. While it is not currently used in any existing superconducting magnets, it is being considered for future use.

3.3.4 Glass production

What is called the Float process or sometimes the Pilkington process is a means by which flat sheets of glass are produced on a large scale. Tin is not required as a component of the glass, but rather is part of the process. Molten tin is the surface upon which the glass is laid out. This surface is more smooth and free of any wrinkles or problems than any solid surface, and thus is well suited for the production of flat pieces of glass. The Pilkington Company website explains it as follows:

"At the heart of the world's glass industry is the float process – invented by Sir Alastair Pilkington in 1952 – which manufactures clear, tinted and coated glass for buildings, and clear and tinted glass for vehicles. The process, originally able to make only 6 mm thick glass, now makes it as thin as 0.4 mm and as thick as 25 mm. Molten glass, at approximately 1,000 °C, is poured continuously from a furnace onto a shallow bath of molten tin. It floats on the tin, spreads out and forms a level surface. Thickness is controlled by the speed at which solidifying glass ribbon is drawn off from the bath. After annealing (controlled cooling) the glass emerges as a 'fire' polished product with virtually parallel surfaces [12]."

3.3.5 Tin in polyvinyl chloride

The long-term performance of polyvinyl chloride (PVC) depends on what are called tin stabilizers – tin compounds incorporating methyl, butyl or octyl groups [13]. The tin is able to react with chloride ions during the slow breakdown of the PVC in sunlight or corrosive atmospheres, and prevent its escape from the material, or prevent the formation of HCl and any subsequent off-gassing of it. This use consumes almost 50,000 tons of tin annually. Indeed, market demand for organo–tin stabilizers is large enough that the corporation Shital Industries devotes significant effort to production of a methyl–tin, a butyl–tin, and an octyl–tin stabilizer [14].

3.4 Possible substitutes

Tin's use in containers – tin cans – is perhaps obviously always in competition with other containers, from aluminum cans to plastic bottles. Certain foods, which may degrade when exposed to UV and visible light, are preferentially packaged in opaque containers, tin being one type.

In specific applications, epoxy resins can substitute for tin-based solders, primarily when there is no need for electrical conductivity in the end-use item.

3.5 Recycling

As with many other metals, tin recycling is a long-established segment of the metals recycling industry. In the United States and Europe, the small amounts of tin and solder that are used in electrical connections are not often recycled, because it is not usually economically feasible. In India and the Far East, what is called e-waste is often recycled for its metals, including tin. This is waste hardware from computers and other electronic devices, each of which does contain small amounts of several metals, including tin. Unfortunately, this recycling involves heating the e-waste and melting the metals to separate them from plastic and other parts, and is not well regulated. This lack of regulation in turn means that such recycling is sometimes performed under dangerous conditions.

Bibliography

[1] USGS Mineral Commodity Summaries. Downloaded as: http://minerals.usgs.gov/minerals/pubs/mcs/2015/mcs2015.pdf
[2] Top 10 Tin Producers. Website. (Accessed 13 April 2015 as: http://metals.about.com/od/Top-10-Producers/tp/The-10-Biggest-Tin-Producers-2013.htm).
[3] Yunnan Tin Group. Website. (Accessed 13 April 2015, as: http://en.ytc.cn/).
[4] Malaysia Smelting Corporation. Website. (Accessed 13 April 2015, as: www.msmelt.com/index.htm).
[5] Minsur. Website. (Accessed 13 April 2015, as: www.minsur.com).
[6] Timah. Website. (Accessed 13 April 2015, as: www.timah.com).
[7] Thaiscaro. Website. (Accessed 13 April 2015, as: www.thaisacro.com).
[8] Yunnan Chengfeng. Website. (Accessed 13 April 2015, as: http://en.yhtin.cn/index.html).
[9] Gunagxi China Tin Group. Website. (Accessed 13 April 2015, as: www.china-tin.com).
[10] EM Vinto. Website. (Accessed 13 April 2015, as: www.vinto.gob.bo).
[11] Metallo Chimique. Website. (Accessed 13 April 2015, as: www.metallo.com).
[12] Pilkington. Website. (Accessed 22 July 2015, as: http://www.pilkington.com/pilkington-information/about+pilkington/education/float+process/default.htm).
[13] PVC. Website. (Accessed 23 July 2015, as: http://www.pvc.org/en/p/organotin-stabilisers).
[14] Shital. Website. (Accessed 23 July 2015, as: http://www.shitalind.com/pvc_tin.htm).

4 Zinc

4.1 Introduction

Zinc is another metal, like copper and iron, that has been used and worked since ancient times, but in almost all cases as an alloy with copper, which is considered to be brass even though small amounts of other elements may also be present. While zinc may have been isolated in antiquity, it was first isolated in a reproducible manner in the mid-1700s by heating a zinc ore and charcoal in a closed container, a chemical process not all that different from what is used commercially today.

The top ten corporate producers of zinc worldwide are shown in Table 4.1. Note that these corporations mine other metals and minerals as well.

Table 4.1: Top ten zinc producers worldwide [1–10].

No.	Name	Location	Amt. (1000s of tons)	Other products
1	Korea Zinc Group	S. Korea	1,100	Bismuth, cadmium, copper, gallium, gold, lead, nickel
2	Nyrstar	Switzerland	1,073	Copper, silver, gold
3	Hindustan Zinc	India	764	Lead, silver, cadmium
4	Glencore Xstrata	Switzerland	651	Coal, iron, copper, nickel, platinum group metals
5	Votorantim	Brazil	577	Aluminum, nickel
6	Boliden	Sweden	456	Copper, gold, lead, silver
7	Shaanxi Nonferrous Metals	China	404	Gold, lead
8	Teck	Canada	293	Copper, steelmaking coal
9	China Minmetals Corp.	China	285	Aluminum, antimony, copper, rare earths, tin, tungsten
10	Noranda Income Fund	Canada	270	Zn–Al alloys

4.2 Refining and isolation

The worldwide production of zinc is very high, generally more than ten million tons annually [11]. Its production by nation is shown in Figure 4.1, in thousands of metric tons. It is evident that China is currently the largest producer by more than twice that of Australia or Peru.

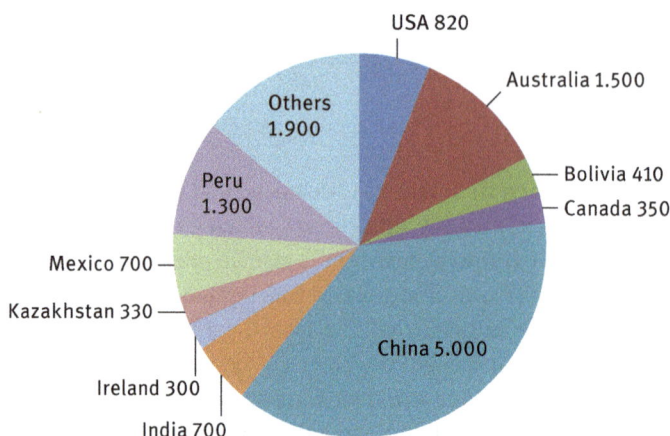

Fig. 4.1: Worldwide zinc production, in thousand of metric tons.

Extraction of zinc from its ores follows several steps that are routine for almost any metal ore. They include:
1. Grinding. This step reduces and makes uniform the size of the ore particles, maximizing the surface area. Usually, particle size is 0.1–1.0 mm.
2. Froth flotation. This step separates and concentrates the mineral ore from what is called gangue, worthless material that coexists in the ore. This step takes advantage of the hydrophobicity or hydrophilicity of different materials in the mix. Surfactants may be required to aid in the separation.
3. Aqueous separation. If necessary, water is removed and the ore further concentrated.
4. Roasting. A large amount of zinc is mined as a sulfide ore, which means the ores must have the sulfur driven off prior to reduction to the metal.

One of the following two steps is then performed on the roasted material, but both are not routinely required:
5. Smelting. This high temperature process requires a carbon source to liberate zinc from zinc oxide.
6. Electrowinning. This step first brings the zinc into solution using sulfuric acid, then deposits it electrolytically onto electrodes, reducing it from solution.

The simplified chemical reactions involved in roasting, smelting, and electrowinning are shown in Scheme 4.1.

When sulfur dioxide is produced in the extraction process, it can be captured for use in producing sulfuric acid, which in turn can be used in the reduction of the zinc. In the case of electrowinning, highly pure zinc can be deposited at the electrode. Interestingly, in the case of smelting, which is a pyrometallurical process, the low boiling

Roasting: $2\,ZnS + 3\,O_{2(g)} \rightarrow 2\,ZnO + 2\,SO_{2(g)}$

Smelting: $C + 2\,ZnO \rightarrow 2\,Zn + CO_2$

and $CO_{(g)} + ZnO \rightarrow Zn + CO_{2(g)}$

Electrowinning: $ZnO + H_2SO_{4(aq)} \rightarrow H_2O + ZnSO_{4(aq)}$

and $2\,ZnSO_{4(aq)} + 2\,H_2O_{(l)} \rightarrow 2\,Zn + O_{2(g)} + 2\,H_2SO_{4(aq)}$

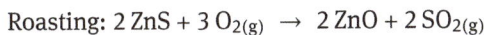

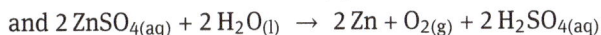

Scheme 4.1: Zinc reduction and isolation.

point of zinc (907 °C) means that zinc in the gaseous state is distilled and then cooled to the solid form.

4.3 Uses

Since the alloys brass and bronze have been discussed in Chapter 3, the other major uses of zinc will be discussed here.

4.3.1 Galvanizing

The USGS Mineral Commodity Summaries comment that galvanizing accounts for 80 % of zinc use [11]. Likewise, Noranda claims at their website: "Its main use is in galvanizing steel for the automotive and construction industries [10]." Galvanizing is a process by which one metal is protected by another. Usually, the more expensive metal, or the metal more critical to an application, is protected by a less expensive metal.

Galvanizing is often done by hot dipping iron, steel, or other metals parts in a bath of molten zinc. The melting point of zinc, 420 °C, is low enough that iron does not appreciably dissolve into the bath.

Because hot dipping applies a relatively thick coat of zinc to the metal surface, when a very thin coat is needed, or when some further protection to a surface will be applied (usually in the form of a paint), it can be done through what is called electro-galvanization. This process, a form of electroplating, requires a zinc salt solution and a zinc anode, as well as the part to be plated as a cathode.

4.3.2 Nickel silver

The term 'nickel silver' refers to a series of alloys, the most common of which is 60 % copper, 20 % zinc and 20 % nickel. Sometimes still called 'German silver' because of where it was first developed and made popular in the nineteenth century, it has been used extensively for cutlery, musical instruments, and jewelry. At what was the height of its use, in the early nineteenth century, German chemist and metallurgist Dr. Lewis Feuchtwanger tried to see it introduced as the alloy of choice for small denomination United States coinage, after he had emigrated to New York City. While he was not successful, his Feuchtwanger cents, examples of which he gave to members of Congress, and which he made into private tokens, stand as an example of a type of coinage that might have been. An example of a Feuchtwanger cent is shown in Figure 4.2 .

Fig. 4.2: Feuchtwanger cent.

4.3.3 Solders

Zinc solders have found use in various electronic applications. They tend to work well in soldering to an aluminum surface, but tend to perform poorly when placed in an environment that will also have sulfur present. As with tin solders discussed in Chapter 3, there is no theory for when and why a particular solder performs well. The basis for their use is always empirical.

4.3.4 Paint pigments

Zinc oxide, ZnO, finds use as a paint pigment, often called zinc white. It creates a flat, opaque white color, and has been found to be preferential to lead white pigment when the painted surface will be in direct sunlight for prolonged periods of time. Lead white tends to discolor over the course of years.

Zinc chromate, $ZnCrO_4$, has been used as a yellow paint, and can be mixed easily with other pigments to produce yellow-greens. It is sometimes still referred to as zinc yellow, even though it is the presence of the chromate ion that imparts the color. This material has been used on aircraft because it prevents oxidation (as most paints do) and also serves as a biocide. Its toxicity, which makes it a good biocide, also means that workers using it must practice strict safety protocols to avoid inhaling any of it.

4.3.5 Sacrificial anodes

In an application that has much in common with tin plating, zinc is sometimes used as what is called a sacrificial anode. A sacrificial anode is made from an active metal, and connected directly to another metal object to protect it from corroding. Ships use sacrificial anodes attached to the hull to prevent and slow corrosion of the hull, while the anode itself dissolves into the surrounding body of water (hence the term 'sacrificial'). There are other instances, usually involving metal parts in harsh environments, that require a sacrificial anode as well.

4.3.6 Batteries

What is often referred to as the silver-oxide battery is actually a galvanic cell that requires two metals (as in most batteries), one being zinc. This is a type of secondary battery, meaning one that can be recharged, and one that remains commercially viable because it has a very high energy per unit volume and energy per unit weight, at least among the batteries that can be classified as aqueous batteries.

While everyday consumer use of these silver-oxide batteries is generally for small applications, in the 1970s NASA chose this type of battery for the lunar rover. Concerning the rover, the NASA website makes the comment: "Power was provided by two 36-V silver–zinc potassium hydroxide non-rechargeable batteries with a capacity of 121 Ah [12]."

The reaction chemistry for this cell is shown in Scheme 4.2. The reaction as shown from left to right is the discharge cycle, and involves the reduction of silver in an alkaline environment, and the oxidation of zinc.

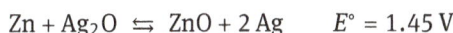

$$Zn + Ag_2O \leftrightharpoons ZnO + 2\,Ag \qquad E° = 1.45\ V$$

Scheme 4.2: Silver oxide battery.

A potassium hydroxide solution must be used to ensure that the reduction and oxidation in the cell occurs.

4.4 Possible substitutes

Aluminum can sometimes be plated onto metal surfaces, as a competitive substitute for zinc galvanizing. Pigments can also be produced using other metal salts, substituting for zinc. However, the continued high production of zinc, and its generally low unit cost (just under US$ 1 per pound) mean that zinc metal and zinc-based materials will be economically favorable materials in many applications.

4.5 Recycling

As with several other metals that are produced in extremely large quantities, such as iron or steel, aluminum, and copper, zinc recycling is a very established, profitable section of the overall recycling industry.

Bibliography

[1]   Korea Zinc Group. Website. (Accessed 20 April 2015, as: www.koreazinc.co.kr).
[2]   Nyrstar. Website. (Accessed 20 April 2015, as: www.nyrstar.com).
[3]   Hindustan Zinc. Website. (Accessed 20 April 2015, as: http://www.hzlindia.com).
[4]   Glencore Xstrata. Website. (Accessed 20 April 2015, as: www.glencore.com).
[5]   Votorantim. Website. (Accessed 20 April 2015, as: www.vmetais.com.br).
[6]   Boliden. Website. (Accessed 20 April 2015, as: www.boliden.com).
[7]   Shaanxi Nonferrous Metals. Website. (Accessed 20 April 2015, as: www.yousegroup.com).
[8]   Teck. Website. (Accessed 20 April 2015, as: www.teck.com).
[9]   China Minmetals Corp. Website. (Accessed 20 April 2015, as: www.minmetals.com).
[10] Noranda Income Fund. Website. (Accessed 20 April 2015, as: www.norandaincomefund.com).
[11] USGS Mineral Commodity Summaries. Downloaded as: http://minerals.usgs.gov/minerals/pubs/mcs/2015/mcs2015.pdf
[12] The Apollo Lunar Roving Vehicle. Website. (Accessed 24 July 2015, as: http://nssdc.gsfc.nasa.gov/planetary/lunar/apollo_lrv.html).

5 Pewter

5.1 Introduction

The term 'pewter' represents a class of alloys that are predominantly tin, usually 85–99 %, and that have a variable composition for the remaining percentage. Antimony, bismuth, copper, lead, and silver have been used in traditional formulations, although lead is seldom used today. Elements like copper and even antimony have been found to produce a harder alloy than pewter formulations lacking these metals. The addition of other metals to the tin has been found to strengthen the resulting material, and also to produce a surface that is pleasing to the eye. Some pewter formulations have an almost blue look to them.

Organizations such as the United States Geological Survey do not track use and production of pewter, although they do track each of the components of the alloy [1]. This is because pewter does not currently have any use that ties directly to national needs, such as defense. There are however companies that still manufacture numerous pewter items [2], and trade organizations that promote the advantages and uses of pewter [3, 4].

5.2 Production

Pewter alloys are produced through direct mixing of the elemental metals. Pewter objects are made through a series of steps, starting with molten tin. Broadly, these are defined as follows by a modern organization, the Worshipful Company of Pewterers:

1. Alloying – alloys are produced in iron pots. Usually tin is added first, both because it is the largest component of the alloy, and because it has a relatively low melting point, 232 °C. Alloys must be mixed thoroughly while molten, and for reproducibility, both exact temperatures and masses of materials are recorded, since the production is a batch process.
2. Casting – the newly alloyed pewter is cast into ingots and allowed to cool.
3. Rolling – the ingots have what are called the upper and lower surface removed, to remove any imperfections that are visible, then they are passed between successive sets of rollers to make them thinner. This process continues with several sets of rollers until the pewter sheet is the desired thickness to work into finished objects.
4. Spinning – the process whereby the flat piece of pewter is lathed into a desired form or shape.
5. Casting – some pewter pieces are made by gravity casting into either rubber or metal molds. Another casting method is the centrifugal method, in which pewter

is poured into a mold that is spun quickly. This latter method can be used to produce hollow pieces that are attractive, but not as heavy as if they were solid.

6. Soldering – the joining of parts that have either been spun or cast into their desired shape.

7. Finishing – as the name of this step implies, it is usually the final step in the process of making an object, and is designed to remove surface irregularities, smooth rough surfaces, and bring out the shine of the metal [4].

5.3 History and traditional uses

Various pewter alloys have been in use in Europe since Roman times. Objects as large as cooking cauldrons have been made from it, although there have been many smaller objects, such as cutlery, made from it as well. When pewter was a relatively rare material, it was also made into patens and chalices for use in churches.

In Renaissance Europe as well as the colonial Americas, a major use of pewter was the production of cooking and eating utensils. Pewter jugs, plates, cutlery, and other kitchen items were common for centuries, as were pewter buttons, candlesticks, and other table setting items, until they were displaced in large part by glass or inexpensive porcelain. Almost all production of pewter prior to the twentieth century was for use in items of this size, simply because a scaled up metals industry was in its infancy, and was at the time focused on iron and steel, as well as copper and its alloys.

5.4 Modern applications

The traditional use of pewterware in churches has declined from the eighteenth century to the present. But pewterware and other pewter household items have remained common, with many of them sold for their decorative appeal, rather than for their functionality. Today pewter tableware is less common than what is generally called 'silverware' made from other metals, but it has not totally disappeared from use. There are other objects made from pewter as well, for use somewhere in households. Figure 5.1 shows an example of a pewter soap dish and ashtray.

5.5 Recycling

There is no large-scale recycling of pewter, essentially because pewter is used only for consumer end-products that are made to last for decades or longer. Pewter tableware and other household items are often passed from owner to owner, often generationally, since they are robust and essentially unbreakable. Additionally, pewter remains quite inexpensive, with no real economic driver for the recycling of the metal.

Fig. 5.1: Modern pewterware.

Bibliography

[1] USGS Mineral Commodity Summaries. Downloaded as: http://minerals.usgs.gov/minerals/pubs/mcs/2015/mcs2015.pdf

[2] ASL Pewter Foundry. Website. (Accessed 24 July 2015, as: http://www.aslpewter.com/index.php).

[3] The Pewter Society. Website. (Accessed 14 April 2015, as: http://www.pewtersociety.org/).

[4] The Worshipful Company of Pewterers. Website. (Accessed 24 July 2015, as: http://www.pewterers.org.uk/pewter/manufacture.html).

6 Gold

6.1 Introduction

No other metal quite piques the interest and the desire of people like gold. Of all the metals on the periodic table, only gold and copper have a color that is not generally thought of as silver. Both have been found as elemental metals in nature. Gold has been found in many river beds, as well as in quartz veins throughout the world.

In what is now Turkey, in ancient times gold nuggets were found as placers in the rivers, sometimes naturally alloyed with silver in what is called electrum. These lumps were valued for their beauty and color, and to verify their weight and value, were often stamped with some form of punch. Eventually these stamps took on specific images, such as the head of a bull or a lion, with the results being the world's first coins – Lydian staters.

Gold was used extensively as coinage systems developed, but gold has also been used as adornment in all cultures that possess it. Cultures as diverse as those of ancient Minos, ancient China, Sumer, the Roman Empire, and pre-Columbian Meso-American and Amazonian peoples all used gold as a form of personal adornment.

The lure of gold has been one of the driving forces for people to move or be moved to parts of the world far from their home. The Spanish and Portuguese claims and conquests to large areas of the New World are in part based on the belief in the presence of gold in those areas. The myth and legend of El Dorado has become part of that belief in vast amounts of undiscovered gold somewhere in the hinterlands of North or South America. Additionally, the settlement of parts of Australia was fueled by gold finds. Likewise, there have been gold rushes to the southern Appalachians in the USA, and the very famous 1849 gold rush to the American West. The Klondike gold rush to the Yukon in 1896 spurred possibly 100,000 people to move there in search of wealth. And more grimly, many who were imprisoned in the Soviet Gulag system during what is now simply called the Terror, in the late 1930s, were sent to the eastern region of Siberia to mine gold.

While these gold rushes are now part of history, even today people are moving back to parts of California, and to the old gold fields, in search of what may have been missed in the past. The search for gold continues [1].

6.2 Refining and isolation

There are hundreds of companies involved in gold mining, production, and refining, simply because of the value of the metal (in 2015 the price was roughly $ 1,200 per troy ounce). Because of its value, many companies that are primarily concerned with the

Table 6.1: Gold-producing companies.

No.	Company name	HQ location	Other products	
1	Barrick Gold Corp.	Canada	Silver	[6]
2	Newmont Mining Corp.	USA	Copper, silver	[7]
3	AngloGold Ashanti	South Africa	Silver	[8]
4	Gold Fields, Ltd.	South Africa	Copper, molybdenum, platinum, palladium, nickel, silver	[9]
5	Newcrest Mining, Ltd.	Australia	Copper	[10]
6	Kinross	Canada	Silver	[11]
7	Goldcorp Inc.	Canada	Silver	[12]
8	Yamana Gold	Canada	Copper, silver	[13]
9	Agnico-Eagle Mines	Canada	Silver, copper, zinc	[14]
10	Polyus Gold	Russia		[15]

mining, production, and refining of other metals also recover small amounts of gold that may exist as co-products extracted from the various ores.

Gold is so universally valued that international and national organizations exist to promote its uses, and it is tracked by national governments [2–5]. Table 6.1 shows a list of the ten largest gold-producing companies, and is limited to this because a full list of all companies would require several pages [6–15]. It should be noted that these companies generally also mine other metals, either as co-products in their gold mining operations, or in separate operations within the administration of the same company.

Traditionally, gold has also been found in what are called nuggets. These placer deposits are usually on the surface or very close to it. The allure of gold is such that

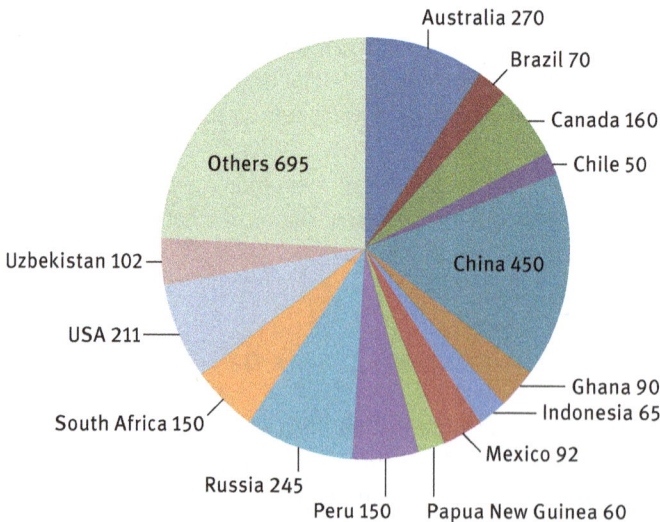

Fig. 6.1: Worldwide gold production, in metric tons.

Table 6.2: Famous gold nuggets.

Name	Weight (oz.t.)	Weight (lb.)	Location of find	Find date, fate
Alaska Centennial	294.1		Ruby, Alaska	1998
Armstrong nugget	80.4		Grant County, Oregon, USA	1913
Beyers and Holtermann	≈ 5,000	630	Hill End, Bathurst, Australia	1872
Boot of Cortez	389		Mexico	1989
Butte nugget	70	6.07	Butte County, California, USA	July 2014, sold to private collector
Dogtown nugget		54	California, USA	1859
Fricot nugget	201	6.25	California Gold Rush	At California State Mining & Mineral Museum
Golden Eagle	1,136		Western Australia	1931
Hand of Faith	875	27	Kingower, Victoria, Australia	1980
Heron	1,008	69.08	Mt. Alexander goldfield	1855
Highland Centennial	27.495		Montana	1989
Ironstone's Crown Jewel	528		Jamestown, California	1992
Lady Hotham		98.5	Ballarat, Australia	Sept 8, 1854
Mojave Gold Nugget	156		Randsburg, California	1977
Normandy nugget	177		Ballarat, Australia	
Pepita Canaa	1,951	133.80	Serra Pelada Mine, Para, Brazil	Sept 13, 1983, now at Banco Central Museum
Welcome Nugget	2,218	152.1	Bakery Hill, Ballarat, Australia	1858, melted in 1859
Welcome Stranger	2,520	173	Moliagul, Victoria, Australia	1869

these nuggets are often given names, some of them quite colorful. A non-exhaustive listing of them is shown in Table 6.2.

These gold nuggets are impressive and continue to fuel the human desire for gold as a source of wealth; but they comprise a very small part of the gold that is extracted and refined by companies each year. Much of current gold production involves chemical separation of extremely small flecks and particles of gold from surrounding rock and soil. Various estimates have been made, for example that for most gold-bearing ores and sands, a box of material one meter square contains only enough gold to make a wedding ring.

Figure 6.1 gives a breakdown of where gold is produced throughout the world. This does not include gold that has been recycled, but rather only gold that has been mined and extracted. Note that this is tracked in metric tons, even though gold is routinely priced in troy ounces when brought to the world's metal markets [2].

The chemical reactions by which gold is extracted from ore can be complex, but a simplified form is shown in Scheme 6.1, in which gold is made soluble as a complex.

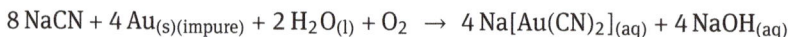

$$8\,NaCN + 4\,Au_{(s)(impure)} + 2\,H_2O_{(l)} + O_2 \rightarrow 4\,Na[Au(CN)_2]_{(aq)} + 4\,NaOH_{(aq)}$$

Scheme 6.1: Gold isolation.

This reaction, sometimes called the Cyanide process, or the MacArthur–Forrest process, must be followed by a means of reducing the gold to its metal form. The technique that accomplishes this is called the Carbon in Pulp process (the CIP process). A summary of its steps are as follows:
1. Carbon particles – washed in a solution with the gold-bearing complex.
2. Removal – carbon-gold complex is separated from the solution and washed.
3. High pH and high temperature – used to remove the gold cyanide from solution.
4. Electro-winning – an electrolysis step that deposits gold metal at the cathode.
5. Smelting – cathodes are smelted for further uses.

This process works particularly well when copper or silver are not co-mingled in the ore batch. It has gained wide use in the past thirty years, and has proven effective when ores must be concentrated from less than 1 % gold to some amount of extractable, economically feasible gold metal.

Anode muds

As mentioned in Chapter 3 on copper, anode muds accumulate in electrolytic processes for reducing highly pure copper. Depending on the batch, these contain varying amounts of silver, gold, and sometimes platinum group metals. The Mineral Commodity Summaries states: "... 7 % of domestic gold was recovered as a byproduct of processing domestic base metals, chiefly copper," concerning gold production in the United States [2].

The Miller process

This process is used to produce highly pure gold from metal batches that have small amounts of other metals in them. A stream of chlorine gas is directed over the metal as it is being heated. The metals in the gold form metal chlorides, leaving pure gold unreacted. The chloride salts are subsequently removed since they are insoluble in the molten gold, leaving gold in purity as high as 99.9 % behind.

The Wohlwill process

This process is an electrolytic one in which gold can be refined to 99.999 % purity, but which requires a solution of chloroauric acid ($HAuCl_4$) that must be maintained and cannot be recovered from the solution. Gold anodes, often called Doré bars, of 95 % purity or greater are used in conjunction with pure gold cathodes in an electrolytic bath. The process deposits only gold at the cathode, which is ultimately removed from the bath as the anode is consumed. The cathode is at 99.999 % purity.

6.3 The carat system – 18, 14, 12, 10 Carat

The purity of gold has been examined, studied, and categorized for centuries, if not for millennia. There are several systems in use today to determine gold purity, with perhaps the easiest one in terms of measurement simply being percent. Additionally, what is called fineness is often expressed in decimal fractions smaller than one. For instance, 99.9 % pure gold is 0.999 fine.

But an older, established system is still used extensively: the carat. This is a means of dividing gold into 24 parts, or levels of purity. Twenty-four carat gold is in theory 100 % pure, 22 carat gold is 91.67 % gold (the alloy used in some coins in the past), 18 carat is 75 % gold, 14 carat gold is 58.33 %, and 10 carat gold is 41.66 % gold.

6.4 Uses

Many people think of gold as a metal to be used for jewelry, while others consider it as a reserve of wealth, in the form of gold ingots and bars. There are certainly more uses than these. The USGS Mineral Commodity Summaries states, concerning gold: "Estimated domestic uses were jewelry, 41 %; electrical and electronics, 35 %; official coins, 18 %; dental, 4 %; and other, 2 % [1]."

6.4.1 Jewelry

Gold jewelry is still often made and valued in terms of the carat system, the older system that divides gold into 24 parts. Pure gold is seldom used in jewelry because it is very soft, and can be easily scratched. Both the 14 carat and 10 carat grades are more often used in jewelry, with the remaining parts being copper, because these alloys are

Fig. 6.2: Examples of college rings in 10 carat gold.

harder, and more resistant to damage and wear. Some 18 carat gold is used in different types of jewelry as well. Two companies, Jostens and Balfour, produce a large number of college, high school, and champion sports rings in the United States, and tend to work in 10 and 14 carat grades, examples of which are shown in Figure 6.2.

6.4.2 Investment coins

In the 1960s, the government of South Africa began the production of one-ounce gold coins called Krugerrands. In more recent years, several national governments have begun producing gold coins that are sometimes called 'bullion coins' and at other times called 'non-circulating legal tender'. A listing of many of these non-circulating

Table 6.3: Gold bullion coins.

Country	Coin Name	Denomination / Size
Australia	Nugget	1/20, 1/10, $\frac{1}{4}$, $\frac{1}{2}$, 1 oz., 2 oz., 10 oz., and 1 kg
Austria	Philharmoniker	1/10, $\frac{1}{4}$, $\frac{1}{2}$, 1 oz.
Canada	Maple Leaf	1 g, 1/20, 1/10, $\frac{1}{4}$, $\frac{1}{2}$, 1 oz.
China	Panda	1/10, $\frac{1}{4}$, $\frac{1}{2}$, 1 oz.
Great Britain	Britannia	1/20, 1/10, $\frac{1}{4}$, $\frac{1}{2}$, 1 oz., 5 oz.
Israel	Tower of David	1 oz.
Kazakhstan	Golden Irbis	1/10, $\frac{1}{4}$, $\frac{1}{2}$, 1 oz.
Malaysia	Kijang Emas	$\frac{1}{4}$, $\frac{1}{2}$, 1 oz.
Mexico	Libertad (Onza)	1/20, 1/10, $\frac{1}{4}$, $\frac{1}{2}$, 1 oz.
Poland	Orzel Bielik	1/10, $\frac{1}{4}$, $\frac{1}{2}$, 1 oz.
South Africa	Krugerrand	1/10, $\frac{1}{4}$, $\frac{1}{2}$, 1 oz.
Switzerland	Vreneli	0.1867 oz.
USA	Eagle	1/10, $\frac{1}{4}$, $\frac{1}{2}$, 1 oz.

legal tender bullion coins are shown in Table 6.3. These bullion coins are produced in addition to the coin programs most of the listed national mints maintain by which special finish gold (and often silver) commemorative coins are marketed to collectors.

The idea behind these bullion coins is to permit smaller investors to own gold, and to do so tangibly, as opposed to buying shares in gold-holding companies, or electronically traded funds. In the last decade, government mints appear to be competing with each other in terms of increasing the fineness of the gold in the coins. Whereas 0.999 fine had been a standard for two decades, more recently, gold coins with 0.9999 fineness have appeared, and even a limited number of 0.99999 fine coins. Also, significantly larger 'coins' have been made, with the current largest being the one tonne gold coin issued by the Perth Mint of Western Australia, with a given face value of $1 million. This appears to be an attempt to surpass the Royal Canadian Mint's (the RCM) 100 kg gold coin of 2007, also with a face value of $1 million.

Interestingly, the RCM claimed that the 100 kg gold coin was something of a marketing idea to promote the fact that it could produce gold at a 0.99999 fineness level, and had planned to make only one. However, shortly after it was created, requests were received from parties who wished to purchase one. To date, the RCM has sold five of these massive gold coins.

6.4.3 Electronics

The use of gold in electronics is almost exclusively for contacts that provide the best electrical transmittance in a circuit. Gold is an excellent conductor, is highly resistant to corrosion, and when the connections are relatively small (often because the device is quite small), the price of the small amount of gold that must be used does not drive up the cost of the end item [3].

6.5 White gold

The term 'white gold' has more than one meaning. In its broadest sense, it means gold alloyed with any other metal to make the overall appearance silver-looking, or white. In a somewhat stricter sense, white gold is 90 % gold and 10 % nickel. This formula has become popular in recent decades because the nickel in the alloy hardens the resulting end-use objects. But the jewelry industry uses the term white gold rather loosely, in that it can mean gold alloyed with palladium, nickel, or even platinum.

6.6 Possible substitutes

The USGS Mineral Commodity Summaries notes that less expensive metals are often clad with gold, not only for jewelry, but for use in electronic components. It also comments that "palladium, platinum, and silver may substitute for gold [1]", but goes no further into detail as to what applications would best accommodate such substitutions.

6.7 Recycling

Gold recycling is a highly developed industry, with everything from gold jewelry to gold in electronics being recycled. But it has been pointed out that even in 2009, when the cost of gold was at record levels, the amount recycled was, "just one percent of the entire aboveground stock [2]".

Bibliography

[1]   Tucker, A. 'Going for the Gold', *Smithsonian*, July–August, 2012, pp. 34–36.
[2]   USGS Mineral Commodity Summaries. Downloaded as: http://minerals.usgs.gov/minerals/pubs/mcs/2015/mcs2015.pdf
[3]   World Gold Council. Website. (Accessed 3 June 2015, as: www.gold.org).
[4]   World Gold Panning Association. Website. (Accessed 3 June 2015, as: http://eldorado2016.com/world-goldpanning-association/).
[5]   Gold Prospectors Association of America. Website. (Accessed 3 June 2015, as: http://www.goldprospectors.org/).
[6]   Barrick Gold Corp. Website. (Accessed 3 June 2015, as: www.barrick.com).
[7]   Newmont Mining Corp. Website. (Accessed 3 June 2015, as: www.newmont.com).
[8]   AngloGold Ashanti. Website. (Accessed 3 June 2015, as: www.anglogoldashanti.com).
[9]   Gold Fields, Ltd. Website. (Accessed 3 June 2015, as: https://www.goldfields.com/).
[10] Newcrest Mining, Ltd. Website. (Accessed 3 June 2015, as: http://www.newcrest.com.au/).
[11] Kinross. Website. (Accessed 3 June 2015, as: http://www.kinross.com/?gclid=CPGeh-qV9MUCFZWCaQodPUMA7w).
[12] Gold Corp. Website. (Accessed 3 June 2015, as: http://www.goldcorp.com/).
[13] Yamana. Website. (Accessed 3 June 2015, as: www.yamana.com).
[14] Agnico-Eagle. Website. (Accessed 3 June 2015, as: www.agnicoeagle.com).
[15] Polyus Gold. Website. (Accessed 3 June 2015, as: www.polyusgold.com).

7 Silver

7.1 Introduction

Silver has as long and as interesting a history as gold, complete with wars being funded by it, explorers conquering new lands for it, and new refining processes being developed for it. Silver has been used as adornment for millennia, with tools and weapons found in some archaeological sites in some way covered or highlighted with silver as a means of emphasizing the wealth and importance of the owner.

The discovery of vast amounts of silver in the New World, specifically in Mexico and Bolivia, ranks as one of the biggest changes in the world's economies [1]. It has recently been proven that by the year 1580, silver from the New World displaced that which had been mined in Europe before it [2]. Since silver has been used for millennia alongside gold for coinage, as nations arose from feudal states, national monetary systems developed that were oftentimes tied to the value of silver.

There are a large number of companies that mine and refine silver, many of which also refine one or more other metals. The world's top ten silver producers are therefore involved in the production of other materials as well as this metal. Table 7.1 lists the top ten producers of silver. Although none of these companies are headquartered in the United States, silver is still mined in Alaska and Nevada, and is a secondary material produced by several other companies within the US.

Silver has been mined extensively throughout the world throughout history, and a breakdown of current silver production is shown in Figure 7.1 [13].

Table 7.1: Top ten world producers of silver [3–12].

No.	Name	Location	Production (M ounces)	Other products
1	Fresnillo plc	Mexico	38.8	Gold
2	BHP Billiton plc	Australia	37.6	Aluminum, copper, iron, manganese, nickel, titanium, uranium
3	KGHM Polska Miedz S.A.	Poland	37.3	Copper
4	GlencoreXstrata plc	Switzerland	37.1	Alumina, copper, ferrochrome, nickel, zinc
5	Goldcorp Inc.	Canada	30.3	Gold, lead, zinc
6	Polymetal International plc	Russia	27.2	
7	Pan American Silver Corp.	Canada	26.0	Gold
8	Volcan Compania Minera S.A.A.	Peru	20.7	Copper, lead, zinc
9	Compania de Minas Buenaventura S.A.A.	Peru	18.9	Lead, zinc
10	Coeur Mining	USA/Mexico	18.0	Gold

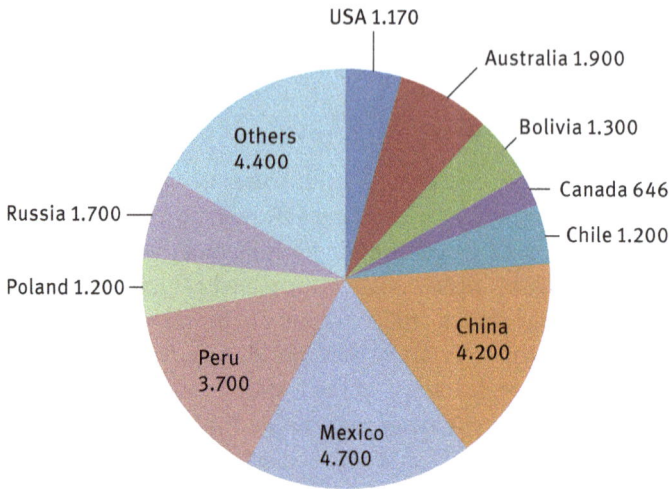

Fig. 7.1: Silver mining, in metric tons.

The percentage listed simply as 'Other' includes mines in areas that throughout history have produced silver on a significant scale, but that now co-produce it with other metals. Greece, Spain, and England for example were major sources of silver during the rise and height of the Roman Empire [1]. One can see that presently, the output of China, Peru and Mexico constitutes almost one half of all the world's production.

7.2 Refining and isolation

While silver has sometimes been found in rather concentrated deposits, more often today it is found in low concentrations within a geologic formation, almost always as a sulfide mineral, and must be concentrated significantly before it is refined. The Washoe process for silver extraction was invented and honed while the Comstock Lode in present-day Nevada was being mined. It is broadly an adaptation of an older, slower process – sometimes called the Patio process or Cazo process – that had been used at Cerro Rico in Bolivia, and elsewhere, for centuries. Reaction chemistry is difficult to show for this, but the steps are as follows:

1. Ore is crushed to the grain size of sand or smaller.
2. Batches of approximately 1,200 pounds are placed in copper pans and made into a slurry.
3. Mercury is added to make an amalgam.
4. Copper sulfate and sodium chloride must be added to the predominantly liquid mixture.

5. Iron paddles or plates are used to stir and agitate the batch.
6. The slurry is heated with steam.
7. Impurities are separated, leaving the silver–mercury amalgam.

It is the mercury that is driven off when the amalgam is separated that is a concern, and which is a problem in this process. The Washoe process has been displaced in modern times. Now, most silver is refined by smelting and leaching processes, followed by electrolytic refining. For most ore batches, especially copper ores, some combination of the following steps is required to extract the silver:

1. Crushing and milling. This brings the ore to a uniform size with high surface area.
2. Flotation separation. This utilizes surfactants to help separate silver minerals from others based on differences in their hydrophobicity. It allows the silver-bearing ore to be both separated and concentrated.
3. From copper ores, smelting of the overall sulfide, followed by electrolytic refining, concentrates the silver in an anode mud or anode slime.
4. The anode mud is further smelted to oxidize all metals except the silver, plus gold and platinum group metals (PGM). This forms a relatively pure ingot.
5. The ingots then undergo electrolysis in a solution that is of both silver nitrate and copper nitrate. This results in high purity silver.

Another method of silver extraction, in this case from lead metal, is what is called the Parkes process. In this, liquid lead metal and liquid zinc are mixed. The two metals when molten are immiscible in each other, but silver impurities in the lead are highly miscible in the zinc. Thus, silver migrates to liquid zinc, leaving pure lead behind. The zinc–silver alloy is then heated in a retort until the zinc is a vapor. The silver is recovered, and the zinc is recaptured. Scheme 7.1 illustrates this in terms of reaction chemistry.

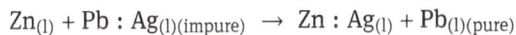

$$Zn_{(l)} + Pb : Ag_{(l)(impure)} \rightarrow Zn : Ag_{(l)} + Pb_{(l)(pure)}$$

Scheme 7.1: Parkes process.

7.3 Uses

7.3.1 Jewelry

Silver jewelry has an ancient history, and silver continues to be a metal that is widely used and prized in jewelry today. A recent newsletter of The Silver Institute claims that the United States is the largest importer of silver jewelry, and that demand in 2015 has risen [1]. The Institute also tracks the use of silver in jewelry on world markets, as well as the rise or fall in consumption based on consumer demands.

7.3.2 Silverware

The past three centuries have seen the rise of silver metal used in what is now ubiquitously called silverware. Prior to the Renaissance, forks were not used at most meals, although knives and spoons have been found in ancient burial sites. With the rise of an affluent class in Europe during the Renaissance, tableware made from silver became a mark of status, and thus a desirable commodity. As silver standards were established, sterling silver came to mean silver of a 92.5 % purity. The remaining metal is routinely copper, although other metals can be used.

7.3.3 Investment coins

Much like the gold bullion coins discussed in Chapter 6, silver bullion coins have been minted by several countries for the past three decades, usually in one-ounce sizes. Most of the countries listed in Table 6.3 also produce silver bullion coins; some now do so in sizes greater than one ounce. The Chinese silver Pandas, for example, are made in 1-oz., 5-oz., 12-oz. and even 1-kg sizes.

The South African Krugerrand remains exclusively a gold bullion coin, but almost all other countries produce a silver version of the bullion coin that represents their nation. Figure 7.2 shows a United States one-ounce silver Eagle.

7.3.4 Photography

Digital photography and laser printing have made significant inroads as a form of competition for silver in photographs, but silver is still used in this application. Silver emulsions form the basis of what becomes the black and white image on paper.

Fig. 7.2: United States silver Eagle bullion coin.

7.3.5 Batteries

The silver oxide battery has already been discussed in Chapter 4, because zinc is used as part of the redox couple. The reaction chemistry for it when run in an alkaline environment (a NaOH or KOH solution) is repeated here, in Scheme 7.2.

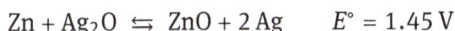

$$Zn + Ag_2O \leftrightarrows ZnO + 2\,Ag \qquad E° = 1.45\,V$$

Scheme 7.2: Silver oxide battery.

Silver can also be used in a battery with an iron couple. Currently, silver–iron batteries are considered somewhat costly, but are finding use in submersibles and telecommunications applications. The cell reaction for this is shown in Scheme 7.3.

$$H_2O + Fe + AgO \leftrightarrows Ag + Fe(OH)_2 \qquad E° = 1.34\,V$$

Scheme 7.3: Silver–iron battery chemistry.

It is evident that the cell potential (the $E°$) is very close for each of these.

7.3.6 Electronics

Silver has found uses in what are termed flexible electronics. Flexible displays utilize silver in their construction, and this use is expected to increase in the future.

Additionally, silver is used in some light emitting diodes (LEDs), in either what is called the adhesion layer or the reflective layer. Additionally, silver wires are routinely used as the bonding wire, to bond to the LED substrate.

Silver is also used in the production of photovoltaic cells (PVs), as well as in multilayer capacitors, usually ceramic capacitors.

We have mentioned various solders in earlier chapters. Silver is used in some solders, although the term is often associated with tin or lead alloys.

7.3.7 Ethylene oxide

Ethylene oxide (EO) is the precursor to the plastic polyethylene oxide (PEO). The large-scale production of EO can be effected through several different synthetic routes, but often requires a silver-containing catalyst to speed the reaction. The direct oxidation of ethylene with diatomic oxygen requires a silver catalyst to make the reaction industrially viable.

7.4 Possible substitutes

As mentioned, laser printing and digital photography have made inroads against silver salts in photography. In most other cases where silver has been displaced for a specific use, it is because the replacement metal or material is less expensive. A perhaps obvious example of this is silverware. A great deal of cutlery is now made from metals such as stainless steel, or from plastic.

7.5 Recycling

Silver is a valuable enough metal that it is recycled on a wide scale. Not only silver jewelry and household objects are recycled. When the price of silver spikes on world markets, even old photographs are brought in for recycling, because traditionally the images were created using silver chloride.

Bibliography

[1] The Silver Institute. Website. (Accessed 26 July 2015, as: https://www.silverinstitute.org/site/silver-essentials/silver-in-history/).
[2] Desaulty AM, Telouk P, Albalat E, Albarede F. Isotopic Ag-Cu-Pb record of silver circulation through 16th–18th century Spain. Proceedings of the National Academies of Science, 2011 (http://www.pnas.org/content/108/22/9002.full).
[3] Fresnillo. Website. (Accessed 20 April 2015, as: www.fresnilloplc.com).
[4] Billiton. Website. (Accessed 20 April 2015, as: www.bhpbilliton.com).
[5] KGHM Polska Miedz. Website. (Accessed 20 April 2015, as: www.kghm.pl).
[6] GlencorXstrata. Website. (Accessed 20 April 2015, as: www.glencore.com).
[7] Goldcorp. Website. (Accessed 20 April 2015, as: www.goldcorp.com).
[8] Polymetal International plc. Website. (Accessed 20 April 2015, as: www.polymetalinternational.com).
[9] Pan American Silver Corp. Website. (Accessed 20 April 2015, as: www.panamericansilver.com).
[10] Volcan Compania Minera S.A.A. Website. (Accessed 20 April 2015, as: www.volcan.com.pe).
[11] Compania de Minas Buenaventura S.A.A. Website. (Accessed 20 April 2015, as: www.buenaventura.com/).
[12] Coeur Mining. Website. (Accessed 20 April 2015, as: www.coeur.com).
[13] USGS Mineral Commodity Summaries. (Downloaded as: http://minerals.usgs.gov/minerals/pubs/mcs/2015/mcs2015.pdf).
[14] The Silver Institute. (https://www.silverinstitute.org/site/wp-content/uploads/2014/07/NEEUof-Silver2014.pdf).

8 Iron and steel

8.1 Introduction

Iron is another metal that has been worked by various peoples throughout the world since ancient times, and one of a very few materials for which ages are often named (there is often a Stone Age, Bronze Age, and Iron Age in a civilization, in that order). It is believed that in ancient times, the decline of the Bronze Age Minoan culture corresponds to the rise of iron ages among people with whom the Minoans traded. Ancient Egypt saw a small amount of iron use, even though the metal is not available near the Nile River. Rather, the source of this ancient iron was meteorites that were recovered and worked by craftsmen [1]. Although iron use appears to have started in the Middle East, it spread relatively quickly into India and to Europe. In the west, the Roman Empire ultimately produced iron for construction, weapons, tools, on what was a very large scale for the time. A study of Greenland ice cores showed that pollution from the Roman working of iron actually polluted areas as far away from the empire as Greenland [2]. The Industrial Revolution however, marks the advent of large-scale production of iron in what can be called modern times. The availability of cheap coal and the improvement and enlargement of furnace technologies allowed the production of iron, especially in Great Britain, to be scaled up far beyond what had been possible in earlier times.

A large number of major companies produce iron and steel; many are multinational, global corporations. Many iron and steel producers in the world also have in-

Table 8.1: Top ten world producers of iron and steel [3–12].

No.	Name	Location	Sales (US$, billions)	Other products
1	Arcelor Mittal	Luxembourg	79.4	
2	Posco	South Korea	56.5	
3	Nippon Steel and Sumitomo Metal	Japan	54.8	Ceramics, chemicals, non-ferrous metals
4	Vale	Brazil	47.6	Fertilizers, non-ferrous metals
5	JFE Holdings	Tokyo, Japan	36.1	
6	Baoshan Iron and Steel	Shanghai, China	30.8	
7	Tata Steel	India	23.8	
8	Nucor	USA, North Carolina	19.1	Non-ferrous metals
9	Metalurgica Gerdau	Brazil	18.5	
10	Kobe Steel, Ltd.	Japan	18.1	Titanium alloys, aluminum, copper, slag products, construction machinery, electronic materials

terests in other chemicals, materials, and raw or finished products, although some concentrate only on products made from iron and steel. The top ten iron and steel producers are listed in Table 8.1.

It can be seen from Table 8.1 that companies may have interests in one nation, but headquarters in another. The first and largest company, Archelor Mittal, certainly does not have all its ore producing operations in the small Grand Duchy of Luxembourg.

The USGS tracks the production of iron by nation in three forms: iron and steel, iron ore, and iron oxide pigments [13]. Figure 8.1 shows iron ore production by country. Note that it is recorded in millions of metric tons, since the industry is a very large, productive one, and since steel is used in so many applications.

Figure 8.1 makes it apparent that China, Australia, and Brazil are currently the largest producers of iron ore. But this does not automatically translate into production of steel, as ore is often shipped to furnaces and refineries, possibly over national borders. Iron and steel production is tracked by several organizations, including: the USGS Mineral Commodity Summaries [13], the World Steel Association [14], and several other professional trade organizations [15–28]. Iron and steel production is shown in Figure 8.2, once again in millions of metric tons.

Figure 8.2 makes it apparent that China currently leads the world in iron production, with over half of the current operations. This production involves the reduction of the iron ore, most often an oxide ore, to the reduced elemental metal, concurrently with trapping of the oxygen, usually by some form of carbon. Unfortunately, there is as yet no economically feasible way to capture the emitted carbon monoxide or carbon dioxide that is produced on such a large scale, which means that production of iron and steel continue to be concurrent with large-scale CO_2 emissions.

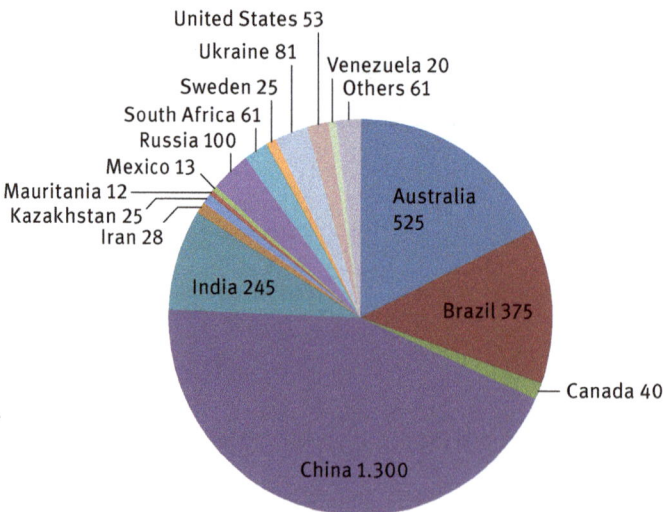

Fig. 8.1: Iron ore production, in millions of metric tons.

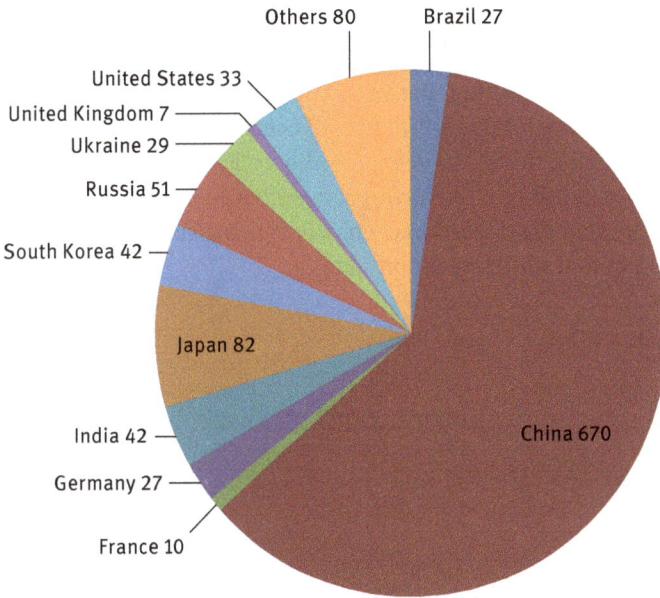

Fig. 8.2: Iron and steel production.

8.2 Ore sources

A variety of iron ores are found widely throughout the world. Table 8.2 lists the major ones, but in is important to mention that most iron is refined from hematite, magnetite, or taconite. It is evident that two of these have very high iron percentages.

Table 8.2: Iron ores.

Name	Formula	Percent iron	Geographic location	Other metals in ore
Ankerite	$Ca(Mg, Mn, Fe)(CO_3)_2$	Varies	Peru	Magnesium, manganese
Goethite	$FeO(OH)$	62.8		
Greenalite	$Fe_4Si_2O_5(OH)_4$	52.3	USA, Minnesota	Both Fe^{2+} and Fe^{3+}
Grunerite	$Fe_7Si_8O_{22}(OH)_2$	39.1	South Africa	
Hematite	Fe_2O_3	69.9		
Laterite	Mixed $Fe_xAl_yO_z$	Varies	India, Australia,	Aluminum, nickel
Limonite	$FeO(OH) \cdot nH_2O$	52.3		
Magnetite	Fe_3O_4	72.4		
Minnesotaite	$(Fe, Mg)_3Si_4O_{10}(OH)_2$	30.7	USA, Minnesota	
Siderite	$FeCO_3$	48.2		
Taconite	Fe_3O_4 mixed with quartz	*Usually > 15	USA, Minnesota, Michigan,	

* Iron in taconite is usually present as dispersed magnetite.

Taconite ores were not mined extensively prior to World War II, since hematite and magnetite ores were easily available. As easy-to-find sources were exhausted, taconite, which is still readily available and plentiful has become a profitable iron source.

The production of iron from natural ores is a series of chemical reductions of some iron compound, usually an oxide such as those in taconite, to the reduced iron metal, and at the same time the removal of oxygen with some other reducing agent, usually carbon monoxide. The simplified reaction chemistry is represented as shown in Scheme 8.1, starting with iron in a high oxidation state.

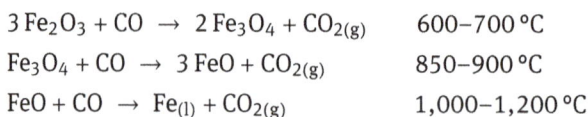

$$3\,Fe_2O_3 + CO \rightarrow 2\,Fe_3O_4 + CO_{2(g)} \qquad 600\text{--}700\,°C$$
$$Fe_3O_4 + CO \rightarrow 3\,FeO + CO_{2(g)} \qquad 850\text{--}900\,°C$$
$$FeO + CO \rightarrow Fe_{(l)} + CO_{2(g)} \qquad 1{,}000\text{--}1{,}200\,°C$$

Scheme 8.1: Reaction chemistry of iron production.

It can be seen readily that at each step, CO functions as the reducing agent, being oxidized to CO_2. Carbon monoxide is added to the system as coke, a more reduced form of carbon, and combusted with air blasts. The basic reaction for this is:

$$2\,C + O_{2(g)} \rightarrow 2\,CO_{(g)} \qquad 200\text{--}700\,°C$$

Because the amount of iron refined is so large, an equally large amount of coke is required for these operations. Additionally, limestone must be used in the reduction of iron ores, since virtually none exist that do not also contain some silicates. Limestone is heated – calcined – to form calcium oxide, then combined with silicates to form the co-product called slag.

As an indicator of how very large the iron and steel industry is, there is actually a National Slag Association, a trade association that concerns itself with uses and potential uses of this co-product material that has been considered an unwanted by-product since the beginning of large-scale iron refining [29]. The simplified reactions for the production of calcium silicate, the main component of slag, are as follows:

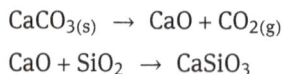

$$CaCO_{3(s)} \rightarrow CaO + CO_{2(g)}$$
$$CaO + SiO_2 \rightarrow CaSiO_3$$

Much of large scale, modern iron production involves the use of blast furnaces. Nearly fifty companies produce steel, and a furnace can produce more than 80,000 short tons of iron each week. A blast furnace diagram is shown in Figure 8.3.

A blast furnace must be lined with brick, often called refractory material, and have an entrance port for iron ore, limestone, and coke at the top. Hot air is blown up through the bottom of the furnace, and it may take hours for the solid material added to the top to travel to the bottom, increasingly hot, zones. During this drop, the

ore, coke,
limestone in

hot waste
gases out

250°C

refractory
lining

hot air in

2000°C

slag out

slag

Fe

Fe out

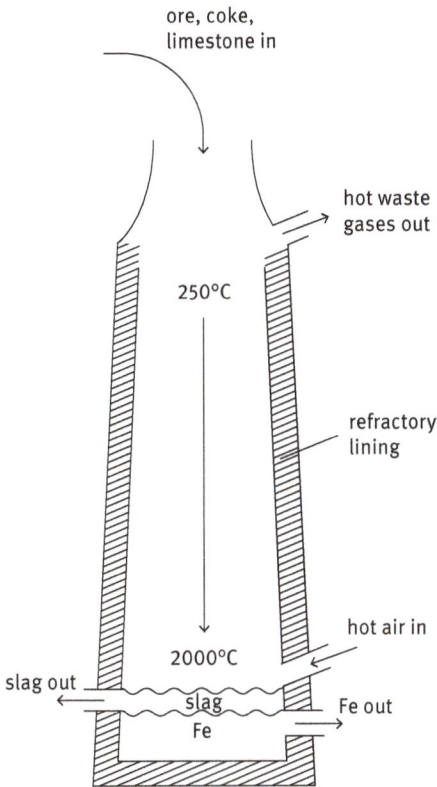

Fig. 8.3: Blast Furnace.

iron ore is reduced to liquid iron, and liquid slag is produced as well. Both must be drained out of the furnace at some interval.

It is difficult to grasp the size of a blast furnace, either from diagrams or photographs. But such operations are routinely large enough that the decision to commission or decommission a blast furnace is a corporate one, made after a great deal of study at the highest levels of a company. Once operational, blast furnaces usually function for years or decades, with stoppages only for maintenance, or in the event of an accident.

8.3 Steel production

The term 'steel' encompasses a large number of alloys of iron and at least one other element. Almost 98 % or refined iron is further utilized for the production of some steel alloy, always based on a specific, desired property.

The World Steel Organization is a trade association that tracks the various types of steel and their uses. The organization states at their website that over 3,500 different

Table 8.3: Common types of steel.

Steel type	Alloying element(s)	Uses or physical properties
Carbon steel	C	Hardness
High speed steel	W	Enhanced hardness
High strength low alloy steel	1.5 % Mn	Enhanced strength
Low alloy	Mn, Cr, < 10 %	Enhanced hardness
Manganese steel	Mn, 12 %	Wear resistance and durability
Stainless steel	Cr 11 %, Ni	Resistance to corrosion
Tool steel	W, Co	Enhanced hardness, drills & cutting tools

steel alloys exist, with a few being rare, and used in small market niches. Table 8.3 is a non-exhaustive list of steels that are used in large amounts.

It is noteworthy that most of the alloys are used to make iron harder. Few elemental materials are made harder by the addition of impurities, and thus in this regard, the steel alloys seem to be uncommon.

The basic oxygen furnace (BOF) and the electric arc furnace (EAF) are the two major ways by which steel is produced. Figure 8.4 shows a diagram of a basic oxygen furnace.

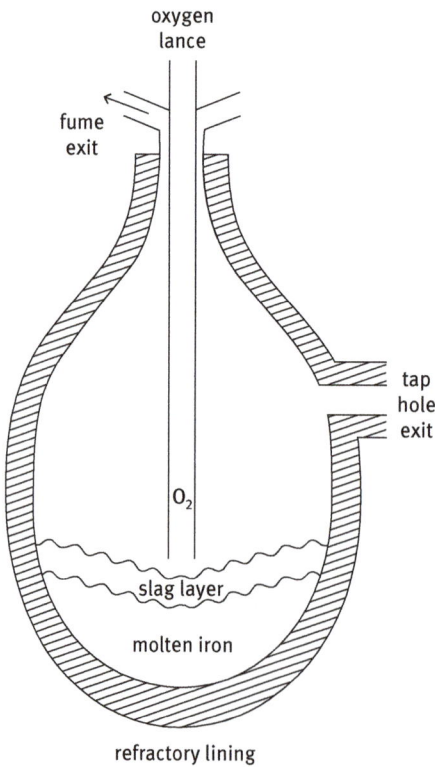

Fig. 8.4: Basic oxygen furnace.

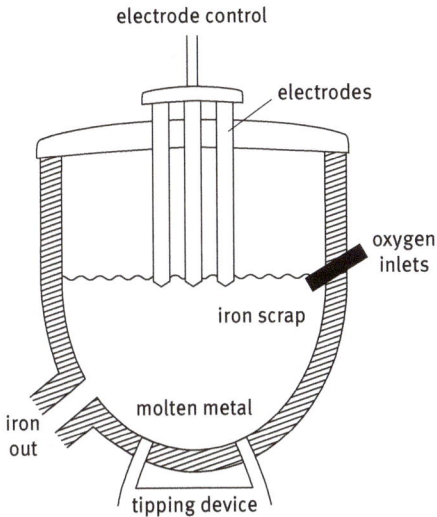

Fig. 8.5: Electric arc furnace.

An oxygen lance is a vital part of any BOF furnace. By introducing high purity oxygen at high flow, carbon in the molten metal is oxidized to CO_2, producing a low-carbon steel. The reaction takes place at hot enough temperatures that it is run in a refractory-lined container, itself made of steel.

Of all the industrial-scale apparatuses for the production of steel, the electric arc furnace is one of the newest. Electric arc furnaces produce steel in batches, which are then poured into forms for further working and shaping. A simplified schematic design is shown in Figure 8.5. The base of any such unit must incorporate a lever system to tip the furnace, enabling the batch of molten metal to be poured.

Scrap iron can be used as the principle feed stock for an electric arc furnace. This is in contrast to the basic oxygen furnace, which more often uses iron ore, coke and limestone [6].

Historically

Throughout history, there have been several different steel formulations that have achieved some fame or notoriety, based on how well materials made from such steels perform, and not necessarily on the manufacturing technique. Three important ones are:

1. Wootz steel or Damascus steel
    This is a type of steel that appears to have depended on small amounts of vanadium for the beautiful, whirled patterns in the finished products, usually swords or knives. Such blades were made in or near Damascus, and thus have this name. But the iron pigs that were the starting metal ultimately came from India. What

was called the secret of Damascus steel was lost for centuries, apparently because the trade routes from India to Syria had been interrupted through some war, plague, or famine, and only in the recent past has the formula been rediscovered [30].

2. Toledo steel

Another city famed for its steel is Toledo, Spain. Much like Damascus steel, Toledo steel was made famous by the swordsmiths who were able to produce beautiful, functional weapons that kept an edge well, and that were lethal in the hands of trained soldiers.

3. Katana

The Japanese long sword, called the katana, used by the traditional samurai class, is another example of steel that has acquired an almost mythical status. Each sword required a minimum of half a year for forging, at least when done in the traditional manner, and it is believed that the process imparts whatever 'impurities' are necessary for the resulting blade to take and hold its extremely sharp edge.

In all these cases, steel manufacturing – or sword manufacturing – never rose above what might be called a cottage industry, a small-scale operation. Since the use of swords in warfare was eclipsed by the advent of firearms, there has never been a need for any of these steels to be manufactured on a large scale.

8.4 Uses of iron and steel

The uses of iron and steel are so numerous that it would be a daunting task to simply list them all. It is not an exaggeration to say that modern life could not exist without the iron and steel production we see and take for granted today.

One agency that tracks the production of iron and steel, the USGS, breaks down the uses into several broad categories, commenting: "The distribution of steel shipments was estimated to be warehouses and steel service centers, 26 %; construction, 16 %; transportation (predominantly automotive), 15 %; cans and containers, 3 %; and other, 40 % [13]." The general population is familiar with cars, reinforcing beams in buildings, bridges, and military vehicles being made from steel, but there are obviously far more as well. Iron and steel are indeed produced on such a large scale that there are several national as well as regional trade organizations dedicated to the promotion and regulation of steel, and that list its uses at their websites [15–19].

Another statement from the USGS Mineral Commodity Summaries gives some indication of how widespread the use of iron and steel is. The comment is in regards to the ability to substitute other materials for iron and steel. It states: "Iron is the least expensive and most widely used metal. In most applications, iron and steel compete either with less expensive nonmetallic materials or with more expensive materials that have a performance advantage. Iron and steel compete with lighter materials, such as aluminum and plastics, in the motor vehicle industry; aluminum, concrete and wood in construction; and aluminum, glass, paper, and plastics in containers [13]." The specific note about aluminum and plastics in the production of motor vehicles relates to the apparently never-ending battle between the need for safety, in which steel generally wins, and light-weighting of vehicles for better fuel economy, in which aluminum and high strength plastics generally win.

8.5 By-product production

The two major by-products of the iron industry are gaseous carbon monoxide and solid slag. Because slag is solid, visible, and accumulates locally there exists the already-mentioned major trade association dedicated to finding some form of use for it – which then enables slag to be considered a secondary product, as opposed to a by-product or waste product. Slag has found a major use as a component of different types of concrete [29].

Unlike slag, carbon dioxide is invisible and odorless. Because of this, for decades there were no efforts to examine any means of controlling CO_2 emissions to the atmosphere. While it is now considered a pollutant, the use, containment, or sequestration of it continues to be a major challenge for the industry.

To provide a sense of scale for how much CO_2 is emitted when iron is produced, we can us the overall reduction reaction for iron production, then perform a stoichiometric conversion to CO_2. Scheme 8.2 shows the reaction. Using one ton of iron oxide

$$Fe_2O_{3(s)} + 3\,CO_{(g)} \rightarrow 2\,Fe_{(l)} + 3\,CO_{2(g)}$$

Scheme 8.2: Overall reaction for iron reduction from ore.

as a starting point one can see how much carbon dioxide is released as a by-product:

$$1\,t\,Fe_2O_3 = 2{,}000\,lb \times 453.59\,g/1\,lb = 907{,}180\,g\,Fe_2O_3$$
$$907{,}180\,g\,Fe_2O_3 \times 1\,mol\,Fe_2O_3/159.687\,g = 5{,}681.0\,mol\,Fe_2O_3$$
$$5{,}681.0\,mol\,Fe_2O_3 \times 3\,mol\,CO_2/1\,mol\,Fe_2O_3 = 17{,}043\,mol\,CO_2$$
$$17{,}043\,mol\,CO_2 \times 44.010\,g\,CO_2/1\,mol\,CO_2 = 750{,}061\,g\,CO_2$$

Converting back to tons:

$$750{,}061\,g\,CO_2 \times 1\,lb/453.59\,g = 1{,}653.6\,lb\,CO_2$$
$$1{,}653.6\,lb \times 1\,t/2{,}000\,lb = 0.827\,t\,CO_2$$

For every ton of iron that is manufactured, almost 0.83 tons of CO_2 is co-produced. When this is considered against the enormous volumes of iron produced, as seen in

Scheme 8.2, it becomes obvious that a practical way to diminish the amount of CO_2 co-produced is a major challenge.

There are other by-products in iron and steel production as well. Most ores are not pure iron oxides, and thus any sulfur in iron sulfides must somehow be captured in the refining process, usually as part of the slag. Additionally, any materials that are not gasified in the furnace processes must be captured in some way in the slag.

8.6 Recycling

Iron and steel recycling are not only major, developed industries in and of themselves, in many operations scrapyards may be devoted entirely to iron and steel alloys. Some projects are so large that the metal to be recycled never makes it to a scrapyard, instead going directly to a furnace for reforming. Perhaps the most obvious example of this is the decommissioning of ships, either from merchant marine or military fleets. These are cut apart, and the refined metal is re-melted and re-used.

Even for operations that are much smaller than what is required for ship salvage, iron and steel recycling is developed enough as an industry that trade organizations exist that are devoted to the recycling industry [30, 31]. In every case, recycling is economically cheaper than finding, mining, and refining new iron or steel.

As mentioned in other chapters within this book, there are widespread recycling operations throughout the world for metals such as copper, aluminum, brasses and several other materials. This means that few scrapyards deal exclusively with iron and steel scrap.

Additionally, in many countries, scrapyards exist that deal exclusively with old automobiles. These then are examples of operations in which steel is one major product, but not the only one, as copper, plastics, and other metals must also be recycled.

Bibliography

[1] Rehren T et al. 5,000 years old Egyptian iron beads made from hammered meteoritic iron. *Journal of Archaeological Science*, Volume 40, Issue 12, December 2013, Pages 4785–4792.
[2] Hong S, Candelone J-P, Patterson CC, Boutron CF. "Greenland Ice Evidence of Hemispheric Lead Pollution Two Millennia Ago by Greek and Roman Civilizations", Science, Vol. 265, No. 5180, pp. 1841–1843 (1994).
[3] ArcelorMittal. Website. (Accessed 21 September, 2014, as: http://corporate.arcelormittal.com/).
[4] Posco. Website. (Accessed 30 July 2015, as: http://www.posco.com/homepage/docs/eng3/jsp/s91a0010001i.jsp).
[5] Nippon Steel and Sumitomo Metal. Website. (Accessed 30 July 2015, as: http://www.nssmc.com/en/).
[6] Vale. Website. (Accessed 30 July 2015, as: http://www.vale.com/EN/aboutvale/Pages/default.aspx).

[7]   JFE Holdings. Website. (Accessed 30 July 2015, as: http://www.jfe-holdings.co.jp/en/).

[8]   Baoshan Iron and Steel. Website. (Accessed 30 July 2015, as: http://www.baosteel.com/group_en/).

[9]   Tata Steel. Website. (Accessed 30 July 2015, as: http://www.tatasteel.com/).

[10] Nucor. Website. (Accessed 30 July 2015, as: http://www.nucor.com/).

[11] Metalurgica Gerdau. Website. (Accessed 30 July 2015, as: http://www.gerdau.com/en#).

[12] Kobe Steel, Ltd. Website. (Accessed 30 July 2015, as: http://www.kobelco.co.jp/english/).

[13] USGS Mineral Commodity Summaries, 2013. Website. (Accessed 7 February 2015 as: http://minerals.usgs.gov/minerals/pubs/mcs/2013/mcs2013.pdf).

[14] World Steel Association. Website. (Accessed 9 February 2015, as: http://www.worldsteel.org/).

[15] World Steel Organization. Website. (Accessed 24 September 2014, as: http://www.worldsteel.org/statistics/BFI-production.html).

[16] American Iron and Steel Institute. Website. (Accessed 12 September 2014, as: http://www.steel.org/).

[17] The Iron and Steel Society: A Division of the Institute of Materials, Minerals, and Mining. Website. (Accessed 12 September 2014, as: http://www.iom3.org/iron-and-steel-society/).

[18] The Iron and Steel Institute of Japan. Website. (Accessed 12 September 2014, as: https://www.isij.or.jp/?lang=english).

[19] Japan Iron and Steel Federation. Website. (Accessed 7 February 2015 as: http://www.jisf.or.jp/en/).

[20] Eurofer, the European Steel Association. Website. (Accessed 12 September 2014, as: http://www.eurofer.org/).

[21] European Confederation of Iron and Steel Industries. Website. (Accessed 12 September 2014, as: http://www.eurofound.europa.eu/emcc/content/organisation/eu019o.htm?p1=topic&p2=Skills_Qualifications).

[22] Korea Iron & Steel Association. Website. (Accessed 12 September 2014, as: http://www.kosa.or.kr/sub/eng/about/sub01.jsp).

[23] African Iron and Steel Association. Website. (Accessed 12 September 2014, as: http://www.afristeel.org/).

[24] South African Iron and Steel Institute. Website. (Accessed 12 September 2014, as: http://www.saisi.co.za/).

[25] Australian Steel Institute. Website. (Accessed 7 February 2015 as: http://steel.org.au/).

[26] British Stainless Steel Association. Website. (Accessed 7 February 2015 as: http://www.bssa.org.uk/).

[27] International Nickel Study Group. Website. (Accessed 7 February 2015 as: http://www.insg.org/).

[28] International Stainless Steel Forum. Website. (Accessed 7 February 2015 as: http://www.worldstainless.org/).

[29] National Slag Association. Website. (Accessed 26 August 2015, as: http://www.nationalslag.org/blast-furnace-slag).

[30] Verhoeven JD, Pendray AH, Dauksch WE. "The key role of impurities in ancient Damascus steel blades". *Journal of Metallurgy* **50**(9): 58. (Accessed 5 April 2016, as: http://www.tms.org/pubs/journals/JOM/9809/Verhoeven-9809.html).

[31] Steel Recycling Institute. Website. (Accessed 7 February 2015 as: http://www.recycle-steel.org/).

[32] Steel Recycling Locator. Website. (Accessed 7 February 2015, as: http://www.recycle-steel.org/Recycling%20Resources/Locator.aspx).

9 Platinum group metals

9.1 Introduction

The platinum group metals (PGMs) include the following six elements: ruthenium, rhodium, palladium, osmium, iridium, and platinum. All have relatively recent histories, with platinum being discovered and isolated from South American ores in the eighteenth century, and the others having been discovered even more recently. The discovery of platinum is intertwined with gold mining in colonial South America, and even today, some PGMs are co-extracted with other precious metals.

Since World War II, a significant interest has developed in the use of the PGMs in some catalytic role, although the first large-scale use of one of these six elements as a catalyst appears to go back to the pioneering work of Fritz Haber in making ammonia from nitrogen and hydrogen with a platinum catalyst, before and during World War I [1].

9.2 Sources, refining and isolation

The production of the platinum group metals takes place on all six inhabited continents, largely because all six of the metals are never all co-located in ores, and also because the recovery of small amounts of one or more of the PGMs is profitable even when they are secondary materials in the refining of other, less expensive metals. The United States Geological Survey tracks the worldwide production of the platinum group metals, and does so in terms of kilograms (instead of metric tons, or thousands of metric tons, as it does with several other metals) [2]. Figure 9.1 illustrates this.

The dominance of South Africa in the production of PGMs is because of the location of what is called the Merensky Reef, part of the Bushveld Igneous Complex, a mining region in the country. Its reserves of PGMs are significantly larger than those found in other countries. For example, platinum was initially discovered in gold-bearing ores from South America during the times of Spanish colonization, but an examination of Figure 9.1 does not show any one country from South America listed as a major producer today. Rather, production from this area is now simply listed in the grouping 'Others'.

The top ten corporate producers of platinum are shown in Table 9.1. Platinum is tracked by several organizations simply because of its value [3], and because of its current use as a catalyst in several different processes.

Table 9.2 shows the top ten producers of palladium. Palladium has only recently found uses that make it an economically important element, whereas the just-mentioned platinum has a longer history of large-scale use, mostly in the form of material in catalytic converters in automobiles. One can note quickly that almost all

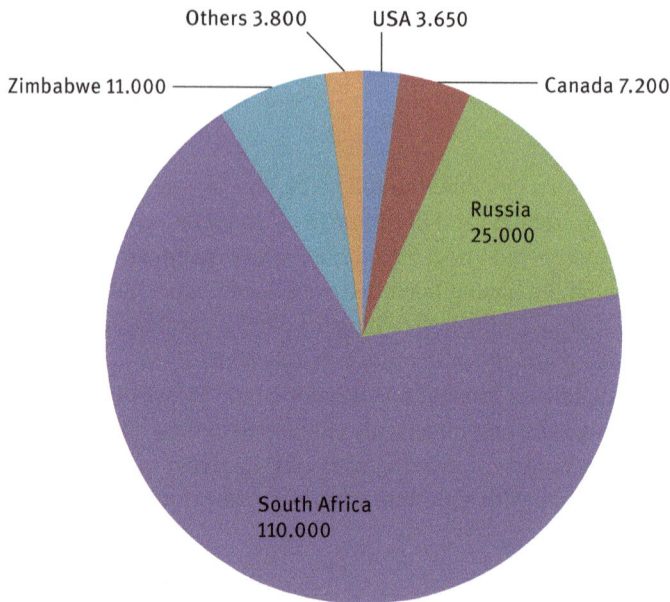

Fig. 9.1: Platinum group metal production, in kilograms.

Table 9.1: Top ten worldwide producers of platinum [4–13].

No.	Name	Location	Amt. (ozs.)	Other products
1	Anglo Platinum	S. Africa, Zimbabwe	2,378,600	
2	Impala Platinum	S. Africa, Zimbabwe	1,582,000	
3	Lonmin Plc	S. Africa	687,372	Copper, nickel, iridium, palladium, rhodium, ruthenium
4	Norilsk Nickel	Russia	683,000	Nickel, palladium
5	Aquarius platinum, Ltd.	S. Africa, Zimbabwe	418,461	Palladium
6	Northam Platinum, Ltd.	S. Africa	175,000	Gold, iridium, palladium, rhodium, ruthenium, silver
7	Stillwater Mining Co.	USA	154,000	Palladium, rhodium
8	Vale SA	Canada	134,000	Gold, iron, nickel, palladium, silver
9	Xstrata	S. Africa, Canada	80,199	Nickel, palladium
10	Asahi Holdings, Inc.	Japan	75,000	

of the firms listed in Table 9.1 also appear in Table 9.2, although the order is different. This is an indicator of how often these two metals appear in nature co-mingled with each other, or co-mingled with other less expensive metals, such as nickel or copper.

Table 9.2: Top ten palladium producers worldwide.

No.	Name	Location	Amt. (M oz)	Other products
1	Norilsk Nickel	Russia	2.731	Nickel, PGMs
2	Anglo Platinum	S. Africa, Zimbabwe	1.395	Platinum
3	Impala Platinum	S. Africa, Zimbabwe	1.020	Platinum
4	Lonmin, Plc.	S. Africa	0.331	Copper, nickel, iridium, palladium, rhodium, ruthenium
5	Stillwater Mining	USA	0.258	Platinum, rhodium
6	Vale SA	Canada	0.251	Gold, iron, nickel, silver, platinum
7	Aquarius Platinum, Ltd.	S. Africa, Zimbabwe	0.238	Platinum
8	North American Palladium	Canada	0.163	
9	Northam Platinum, Ltd.	S. Africa	0.096	Gold, iridium, platinum, rhodium, ruthenium, silver
10	Xstrata	S. Africa, Canada	0.043	Nickel, platinum

The other four PGMs are not produced on nearly as large a scale as platinum and palladium, which is why they are tracked as one commodity inclusive of six elements – PGMs.

The separation of all six platinum group metals shown here, in Scheme 9.1, is somewhat idealized, since all six elements do not usually occur in a single ore. Notice though that the separation reactions generally depend upon the solubility of each element in different acidic or basic environments. Note that the starting material for the

in aqua regia: $PGM_{(s)} \rightarrow (Os, Ru, Rh, Ir, Pt)_{(s)} + (Pd, Au)_{(aq)}$

$NaHSO_{4(l)} + (Os, Ru, Rh, Ir)_{(s)} \rightarrow Rh_2(SO_4)_3 + (Os, Ru, Ir)_{(s)}$

$Na_2O_{2(l)} + (Os, Ru, Ir)_{(s)} \rightarrow Ir_{(ppt)} + (Os, Ru)_{(sol)}$

$(Os, Ru)_{(sol)} + O_2 \rightarrow OsO_4 + RuO_4$

$NH_4Cl + OsO_4/RuO_4 \rightarrow (NH_4)_3RuCl_{6(s)} + OsO_4$

Scheme 9.1: Idealized separation and isolation of the PGMs.

initial separation is often what is called an anode mud or an anode slime, recovered from the large-scale production of a less expensive metal, such as copper or nickel. Each batch of anode mud is different, and thus the separation steps may need to be adjusted. Also, note that in Scheme 9.1 gold is listed in the first reaction. This is because gold is another precious metal often found in anode muds, and co-mingled with different PGMs.

When platinum is recovered from anode muds, it is usually from a copper or nickel operation, and is considered a secondary product. It tends to be co-mingled with palladium, and is separated based on the two metals' solubilities in the acid mixture aqua regia, and the subsequent addition of ammonium chloride.

Scheme 9.1 also shows the addition of molten sodium bisulfate, which is effective at separating rhodium from the other metals. Additionally, the separation of iridium must be followed by its reduction via dissolving in aqua regia, followed by evaporation of the solution and reduction under a hydrogen atmosphere.

9.3 Uses

The platinum group metals have found a wide variety of uses in the industrialized world, although several of them remain small, niche applications. A non-exhaustive listing includes the following:
1. Automotive catalysts for cars and light trucks
2. Chemical process catalysts
3. Catalysts for petroleum refining
4. Laboratory crucibles (predominantly platinum)
5. Computer hard disks (predominantly platinum and ruthenium) [3]
6. Ceramic capacitors
7. Hybridized integrated circuits
8. Dental fillings and restoratives
9. Jewelry
10. Non-circulating legal tender bullion coinage

9.3.1 Ruthenium

There are no large-scale uses of ruthenium metal, and no major uses for ruthenium compounds. However, ruthenium is sometimes alloyed with platinum, palladium, or titanium. Platinum and palladium alloys with ruthenium have found uses in applications where high and low temperatures will be required, and in medical instruments, in the latter case because of their inertness. In the case of alloying ruthenium with titanium, the addition of ruthenium makes the final alloy extremely hard and corrosion resistant. Titanium alloys containing only 0.1 % ruthenium have been found to be many times more corrosion resistant than titanium alone.

9.3.2 Osmium

Osmium is one of the smallest metals markets in the world, with less than 100 kg produced each year. Osmium is always recovered as a co-product of some other metal refining operation, usually nickel refining. Its use in alloys, such as iridium-osmium alloys, results in hard metal materials that have found niche uses in pen nibs, phonograph needles, and other small end-use applications. This is because the end-use item is rather small and the price of the object is thus not prohibitively large.

Osmium also finds use as osmium tetroxide, OsO_4, in a variety of different applications. Two of the now traditional uses are for the cis-oxidation of alkenes, and for biological staining in transmission electron microscopy. The compound can be formed by direct heating of the powdered metal in oxygen gas, usually at 400 °C. It is both volatile and toxic, and thus has been replaced as a choice for alkene oxidations in most cases.

9.3.3 Rhodium

There is very little rhodium produced annually; what is refined is usually an anode mud from nickel refining, or occasionally from platinum refining. The metal is often alloyed with platinum or palladium, because the resulting alloys are hard and corrosion resistant. Specialized spark plug parts, and some high temperature laboratory ware which requires a material be particularly inert, are two applications for such alloys.

9.3.4 Iridium

Iridium is generally considered to be the rarest of the PGMs, and is produced in small quantities each year, usually less than ten tons. Indeed, iridium production is often measured in ounces.

Iridium has what might be called a traditional use, coating the prongs of diamond engagement rings, where it gives the diamond a whiter look. Iridium has found use more recently though, in making crucibles for pure crystal growth, and in some small electronic devices. This has driven the price of the metal upwards on world markets in the recent past.

Iridium compounds do find use in a catalytic role in what is called the Cativa process, a process pioneered by BP in the late 1990s, which produces acetic acid from methanol. Figure 9.2 shows the process, with an emphasis on the role iridium plays in the cycle.

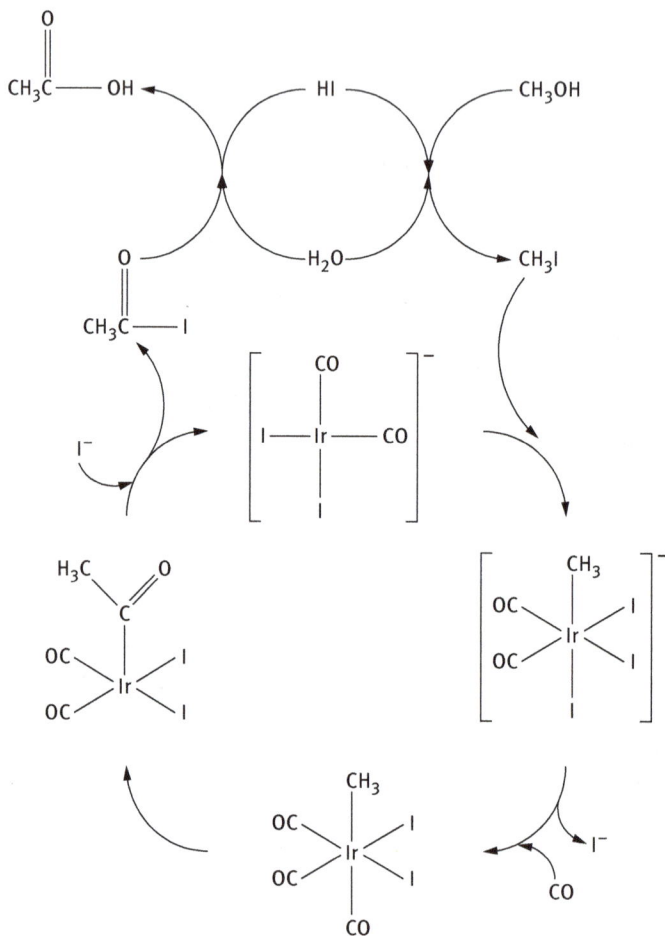

Fig. 9.2: Cativa process for acetic acid production.

9.3.5 Palladium

Palladium has been found to be useful in catalytic converters in automobile exhausts, as has platinum, and thus has become a very useful metal since the phasing out of leaded gasoline. Palladium also finds some use in different jewelry, and has been marketed as palladium commemorative coins by several world mints. This latter use seems to be an outgrowth of the broad commemorative coin programs of many nations, as well as of gold, silver, and platinum bullion coins that have been marketed in the past thirty years, and has not been met with as much collector demand as coins made from the other three precious metals. The Royal Canadian Mint has been able to market a one-ounce palladium bullion coin, but has not done so annually since 2009.

9.3.6 Platinum

Jewelry
Platinum has an established history of use in jewelry. It can be used as the primary metal for different items, or as an alloy. The term 'white gold' is sometimes used for gold alloyed with platinum that has a whitish look, but this term is used loosely, and can be utilized to describe gold that is alloyed with palladium or with nickel.

Catalysts
Automotive catalysts often use platinum as part of the catalytic converter. The platinum is finely dispersed into a ceramic honeycomb structure, to maximize the platinum surface area. Platinum on the honeycomb surface can be used to reduce nitrogen oxides to benign exhaust gases, nitrogen and oxygen. A second and third stage can be used to catalyze the oxidation of carbon monoxide and excess hydrocarbons to carbon dioxide and water. Usually, platinum is used with rhodium as a reduction catalyst, and with palladium can be used as an oxidation catalyst.

Bullion coins
Platinum has taken a place beside gold and silver in terms of the production of platinum coins by various national mints. The idea of offering platinum coins in fractions of an ounce, or in a one-ounce size, is to provide small investors with a means by which they can own the precious metal directly. These coins have not been issued for as long as gold bullion and silver bullion coins. For example, the United States platinum $\frac{1}{10}$ oz., $\frac{1}{4}$ oz., $\frac{1}{2}$ oz. and 1--oz. platinum Eagles were first issued for sale to the public in 1997, while the gold and silver Eagles were first available in 1986. Table 9.3 lists the countries that produce platinum bullion coins.

Platinum compounds
Several platinum compounds have found some use on a relatively large scale. The complex cis-platin, an anti-cancer drug first discovered in 1965, is perhaps the most

Table 9.3: Platinum bullion coins

Nation	Coin name	Weight
Australia	Koala	1/20, 1/10, $\frac{1}{4}$, $\frac{1}{2}$, 1 t oz.
Australia	Platypus	1 t oz.
Canada	Maple leaf	1/20, 1/10, $\frac{1}{4}$, $\frac{1}{2}$, 1 t oz.
Isle of Man	Noble	1/20, 1/10, $\frac{1}{4}$, $\frac{1}{2}$, 1 t oz.
Mexico	Libertad	$\frac{1}{4}$ t oz.
United States	Eagle	1/10, $\frac{1}{4}$, $\frac{1}{2}$, 1 t oz.

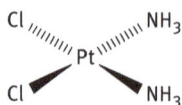

Fig. 9.3: Structure of cis-platin.

well-known example [15] (Figure 9.3). This drug is made by more than a dozen companies today, and has been found to be useful on several different cancers.

9.4 Possible substitutes

The PGMs routinely find industrial uses because of their inert nature, with the exception of platinum and palladium in catalytic converters. Perhaps obviously, in this application they are used precisely because of their reactivity – but also because they are not needed in any large or stoichiometric amount. Since palladium is almost always the less expensive of the two, it is substituted for platinum wherever possible. However, substituting some even less expensive metal for palladium in catalytic converters has not yet been found to be effective.

9.5 Recycling

All of the PGMs are recycled in virtually all parts of the world, because they are so valuable, even in small quantities. Platinum and palladium are recycled from automotive catalytic converters when automobiles are scrapped and their other components are recycled.

Bibliography

[1]   Haber F. US Patent 1202995. Production of Ammonia. Website. (Accessed 6 May 2015 as: http://www.google.com/patents/US1202995).
[2]   USGS Mineral Commodity Summaries. Downloaded as: http://minerals.usgs.gov/minerals/pubs/mcs/2015/mcs2015.pdf
[3]   International Platinum Group Metals Association. Website. (Accessed 23 August 2015, as: http://ipa-news.com/).
[4]   Anglo Platinum. Website. (Accessed 27 April 2015, as: www.angloamerican.com).
[5]   Impala Platinum. Website. (Accessed 27 April 2015, as: www.implats.co.za).
[6]   Lonmin Plc. Website. (Accessed 27 April 2015, as: www.lonmin.com).
[7]   Norilsk Nickel. Website. (Accessed 27 April 2015, as: www.norilsk.ru/en).
[8]   Aquarius Platinum, Ltd. Website. (Accessed 27 April 2015, as: http://aquariusplatinum.com).
[9]   Northam Platinum, Ltd. Website. (Accessed 27 April 2015, as: http://www.northam.co.za).

[10] Stillwater Mining Company. Website. (Accessed 27 April 2015, as: www.stillwatermining.com).

[11] Vale SA. Website. (Accessed 27 April 2015, as: www.vale.com).

[12] Xstrata. Website. (Accessed 27 April 2015, as: www.glencorexstrata.com).

[13] Asahi Holdings, Inc. Website. (Accessed 27 April 2015, as: www.asahiholdings.com).

[14] North American Palladium, Ltd. Website. (Accessed 27 April 2015, as: www.napalladium.com).

[15] The 'Accidental' Cure – Platinum-based Treatment for Cancer: The Discovery of Cisplatin. Website. (Accessed 3 September 2015, as: http://www.cancer.gov/research/progress/discovery/cisplatin).

10 Nickel

10.1 Introduction

Nickel has a relatively short history from discovery to industrial use, having been first considered to be an element only in 1751. After its initial large-scale use, as an alloying element in steel, it has found numerous other uses that make it a metal that has truly transformed the world and several industries. There are now enough applications for nickel that a trade organization exists in part to promote its uses [1].

Table 10.1 lists the top ten producers of nickel worldwide. It can be seen that several of these companies also have interests in other metals, and thus are seen in other chapters within this book as well [2–11].

Interestingly, roughly two-thirds of cobalt production is in conjunction with refining of nickel, although co-production with copper accounts for almost all of the remaining third of production. Table 10.2 lists the top ten cobalt producers. As might be expected, there is some overlap between Tables 10.1 and 10.2 [12–16].

Nickel is mined and refined on all six inhabited continents, and as mentioned, can be the co-product of some other mining operation. The USGS Mineral Commodity Summaries tracks nickel production, because its use is always connected with the needs of various national departments or ministries of defense. Figure 10.1 shows the world production of nickel by country, in metric tons.

Figure 10.1 makes it clear that while some nations have large production capability, there appears to be no single nation that dominates the refining of nickel.

Table 10.1: Top ten nickel producers worldwide.

No.	Name	Location	Amt. (in kt)	Other products
1	MMC Norilsk Nickel	Russia	286	Copper, palladium, platinum
2	Vale SA	Brazil	206	Iron
3	Jinchuan Group Ltd.	China	127	Cobalt, copper, platinum group metals (PGMs)
4	Xstrata Plc		106	PGMs
5	BHP Billiton Ltd.	Australia	83	Aluminum, copper, iron, manganese, titanium, uranium
6	Sumitomo Metal Mining Co.	Japan	65	Copper, gold
7	Eramet SA	France	54	Manganese
8	Anglo American Plc.	England	48	Coal, copper, diamonds, iron, platinum
9	Sherritt International Corp.	Canada	35	
10	Minara Resources Ltd.	Australia	30	Cobalt

Table 10.2: Top ten world producers of cobalt.

No.	Name	Location	Amt. (metric tons)	Other products
1	Collectively: Jinchuan Nonferrous Metals, Huayou Cobalt, Jiangsu Cobalt Nickel Metal, Shenzhen Green Eco-manufacture High-Tech Co.	China	30,200	Nickel
2	OM Group, Inc.	Finland	10,547	
3	Chambishi / ENRC	Zambia	5,435	Copper
4	Umicore	Belgium	4,200	
5	ICCI / Sherritt	Canada	3,792	Nickel
6	Xstrata	Norway	2,969	Nickel
7	Sumitomo	Japan	2,542	Copper, gold, nickel
8	Minara	Australia	2,400	Nickel
9	Queensland Nickel PL	Australia	2,369	Nickel
10	Norilsk Nickel	Russia	2,186	Copper, nickel, palladium, platinum

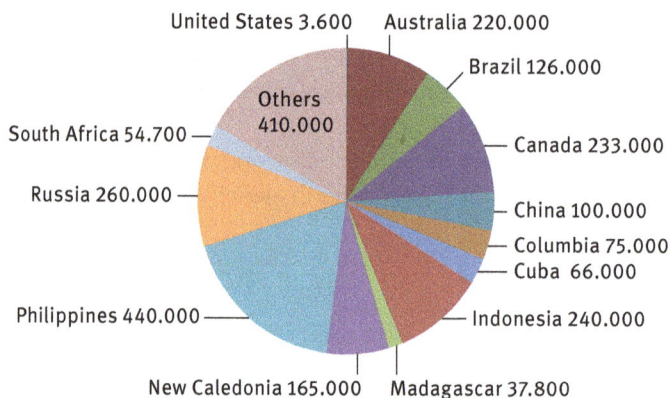

Fig. 10.1: Nickel production by country, in metric tons.

10.2 Refining and isolation

While nickel can be co-produced from metal ores that contain some other element, such as copper, nickel is routinely purified to a high concentration through what is called the Mond process, also sometimes called the Carbonyl process. This process has been used on a large scale for almost 120 years, and takes advantage of the fact that nickel forms volatile carbonyl compounds in a reversible manner, something that appears to be unique to this element. The reaction chemistry, in simplified form, is shown in Scheme 10.1.

$$H_{2(g)} + NiO_{(s)} \rightarrow Ni_{(s)} + H_2O_{(g)} \qquad \text{at } 200\,°C$$
$$Ni_{(s)} + 4\,CO_{(g)} \rightarrow Ni(CO)_{4(g)} \qquad \text{at } 60\,°C$$
$$Ni(CO)_{4(g)} \rightarrow Ni_{(s)} + 4\,CO_{(g)} \qquad \text{at } 220–250\,°C$$

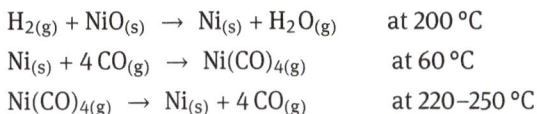

Scheme 10.1: The Mond process.

In the first step, other metals such as cobalt or iron also react, but are not reduced as nickel is. In the third step, it is only nickel that is again reduced to the metal. The solid nickel deposits onto nickel spheres that are in the reaction container.

10.3 Uses

Nickel finds considerable use as an element, but numerous uses in a variety of alloys as well. The USGS Mineral Commodity Summaries states, concerning its uses: "Approximately 45 % of the primary nickel consumed went into stainless and alloy steel production, 43 % into non-ferrous alloys and superalloys, 7 % into electroplating, and 5 % into other uses [17]." Note that 'primary nickel' does not include nickel scrap, which is often in the form of steel that has been slated for recycling.

10.3.1 Steels

The largest use of nickel in any alloy is in steel alloys. What is called stainless steel is almost always 8–12 % nickel, and contains some chromium as well. Different professional organizations have made tables of stainless steels, and given them alphanumeric designations. ASTM specification A240, for example, covers stainless steels that include chromium, nickel, and manganese as alloy components. What is called 304 grade stainless steel – 8 % Ni and 18 % Cr – is used in numerous consumer products such as cutlery and kitchenware. The 316 grade has molybdenum in the alloy as well, but is used where even greater corrosion resistance is required.

10.3.2 Superalloys

The term 'superalloy' has been used rather loosely in common literature in the recent past, but the Nickel Institute takes some pains to define such alloys. Their website states: "The term 'superalloy' is applied to alloys which have outstanding high temperature strength and oxidation resistance. The nickel-based superalloys contain carefully balanced alloying additions of chromium, cobalt, aluminium, titanium and other elements [1]."

Perhaps obviously, the website from such an organization is biased towards their materials, but does spell out what the characteristics of such materials are, even while being less-than-complete about delineating what components are in such alloys.

10.3.3 Shape memory alloys

Shape memory alloys are able to deform and reassume their original shape, usually several hundreds of times. Nickel–titanium alloys have been known to exhibit such macroscopic behavior. Very recently, a four-component alloy has been found that can undergo such deformation more than ten million times [18]. The alloy's stoichiometry gives it the rather unusual formula: $Ti_{54.7}Ni_{30.7}Cu_{12.3}Co_{2.3}$. It is the addition of the small amount of cobalt that appears to produce this ability to deform and reform as many times as it does. While this alloy has not yet been scaled up to industrial size yet, the interest in it and its potential uses is keen.

10.3.4 Plating

Nickel plating is used in numerous applications to produce a metallic product that is corrosion resistant, or wear resistant, even in harsh environments. Additionally, nickel plating tends to give many objects an attractive shine and overall look.

As with other metal plating techniques, nickel plating requires a nickel anode, a solution of a nickel salt (the nickel ions in such a solution are Ni^{2+}), and a metal object to act as the cathode. Application of an electric current both deposits nickel ions at the cathode, and oxidizes nickel from the anode. With enough time, the anode must be replaced.

10.3.5 Nickels

The term 'nickel', meaning a USA five-cent piece, was first introduced in the mid-1800s, as industrialist Mr. Joseph Wharton urged members of the United States Congress to use this metal in various coins (this is the same Joseph Wharton who was the founder of Bethlehem Steel, as well as Swarthmore College, and the Wharton School within the University of Pennsylvania). Prior to the use of nickel metal, five-cent coins in the United States were referred to as half-dimes, and were made of the same 90 % silver, 10 % copper alloys as the other, larger silver coins.

Current United States coinage uses 25 % nickel and 75 % copper in the five-cent coins, as an alloy. In ten-cent, 25-cent and 50-cent pieces, a copper core is clad with a 75 %–25 % copper–nickel layer on each side, giving the coins a white silver look, and an overall composition of 91.67 % copper and 8.33 % nickel. There is no longer any elemental silver in them.

Canada and nickel in coins

Canada, having much more nickel than many other countries, has used it extensively in coinage. The Canadian five-cent piece was made of nickel metal from 1922–1981 with the exception of the years of World War II and the Korean War, when nickel was required for the war effort and war material.

What are called Canadian 'specimen silver' dollars were nickel from 1968–1984. The composition was adopted because of the rising price of silver metal. Starting in 1987, the golden colored Canadian 'loonies' – one-dollar coins so named because of the loon on the reverse – were made of 91.5 % nickel with a bronze plating.

Additionally, Canadian dimes and quarters were made of nickel for decades, until less expensive alloys were adopted.

Nickel in British coins

Nickel metal is currently used in an alloy in all coins of Great Britain with the exception of the one-penny and two-penny pieces. The bimetallic two-pound coin for example has nickel in both rings, although much less in the gold-colored outer ring than in the silver-colored inner ring.

Euro coins

The one-Euro and two-Euro coins are both bimetallic, and both contain nickel in each ring. The gold-colored rings are what is known as nickel-brass, while the silver-colored rings are generally called copper–nickel.

Overall, nickel has found a large use in the coinage of numerous countries, although the use of nickel in various steels remains the largest single application of the metal.

10.4 Possible substitutes

Substitutes for nickel usually come in the form of some other component added to a steel alloy. Nickel is an inexpensive enough metal, and is produced on a large enough scale however, that substitutes are not often sought out, as the substitute usually does not convey any cost saving.

10.5 Recycling

Nickel in steel alloys is almost always recycled, because it is much less expensive to do so than to refine new steel. Nickel for end-use products such as coins is hardly ever recycled, unless the effort is governmentally directed. For example, when the Euro coins were first introduced, national banks recalled their older coins, and when possible, recycled the metal into new coinage, or into other uses.

Bibliography

[1] The Nickel Institute. Website. (Accessed 13 April 2015, as: https://nickelinstitute.org/).
[2] MMC Norilsk Nickel. Website. (Accessed 20 April 2015, as: www.nornik.ru).
[3] Vale SA. Website. (Accessed 20 April 2015, as: www.vale.com).
[4] Jinchuan Group Ltd. Website. (Accessed 20 April 2015, as: www.jnmc.com).
[5] Xstrata Plc. Website. (Accessed 20 April 2015, as: www.xstrata.com).
[6] BHP Billiton Ltd. Website. (Accessed 20 April 2015, as: www.bhpbilliton.com).
[7] Sumitomo Metal. Website. (Accessed 20 April 2015, as: www.smm.co.jp).
[8] Eramet SA. Website. (Accessed 20 April 2015, as: www.eramet.fr/us).
[9] Anglo American Plc. Website. (Accessed 20 April 2015, as: www.angloamerican.com).
[10] Sherritt International Corp. Website. (Accessed 20 April 2015, as: www.sherritt.com).
[11] Minara Resources Ltd. Website. (Accessed 20 April 2015, as: www.minara.com.au).
[12] Jiangsu Cobalt. Website. (Accessed 27 April 2015, as: www.jsklk.com).
[13] Freeport Cobalt. Website. (Accessed 27 April 2015, as: www.freeportcobalt.com).
[14] ERG. Website. (Accessed 27 April 2015, as: www.enrc.com).
[15] Umicore. Website. (Accessed 27 April 2015, as: http://csm.umicore.com).
[16] Queensland Nickel PL. Website. (Accessed 27 April 2015, as: www.qni.com.au).
[17] USGS Mineral Commodity Summaries. Downloaded as: http://minerals.usgs.gov/minerals/pubs/mcs/2015/mcs2015.pdf
[18] Ultralow-fatigue shape memory alloy films. Christoph Chluba, Wenwei Ge, Rodrigo Lima de Miranda, Julian Strobel, Lorenz Kienle, Eckhard Quandt, Manfred Wuttig. Website. (Accessed 8 June 2015, as: http://www.sciencemag.org/content/348/6238/1004.short).

11 Aluminum

11.1 Introduction

Aluminum, first discovered in 1827, was an extremely difficult metal to obtain from its chief ore, bauxite, before the advent of what is now called the Hall–Heroult process, patented in 1887. Thus, for several decades, aluminum was treated as a precious metal, based on its then high price. The Hall–Heroult process brought the price of aluminum down dramatically, but depended on the availability of cheap electricity in the form of the electric dynamo, which was only patented the year before. Since the availability of cheap electricity and the discovery of the Hall–Heroult process, aluminum has become one of the major industrial metals, with numerous uses in a wide variety of applications, and with a very low initial cost. Table 11.1 shows the top ten aluminum producers in the world today.

Figure 11.1 shows the production of aluminum in a different way, by country, in thousands of metric tons. It is evident from the figure that China currently dominates the world production of this metal, although other countries such as Russia and Canada also produce a large amount. But evidence that aluminum is an important commodity today is evidenced by the fact that the United States Geological Survey tracks it, and by the fact that there are national and international associations devoted to its manufacture, promotion, and use [11–18].

Table 11.1: Top ten aluminum producers worldwide [1–10].

No.	Name	Location	Amt. (M of metric tons)	Other products
1	UC Rusal	Russia	4.173	
2	Alcoa, Inc.	USA	3.742	
3	Aluminum Corp. of China	China	3.502	Copper, gallium, rare earths
4	China Power Investment Corp.	China	2.693	Coal, power generation
5	Rio Tinto Alcan Inc.	Canada	2.174	Power generation
6	Norsk Hydro ASA	Norway	1.985	Alumina
7	China Hongqiao Group Ltd.	China	1.821	
8	Shangdong Weiqiao Aluminum & Power Co.	China	1.715	Power generation
9	Shangdong Xinfa Aluminum & Electricity Group Ltd.	China	1.63	Power generation
10	Dubai Aluminum Co.	UAE	1.42	Water desalination

United States 1.720   Argentina 425

Australia 1.680

Bahrain 930

Brazil 960

Canada 2.940

Other 4.440

United Arab Emirates 2.400

S. Africa 735

Saudi Arabia 500

Russia 3.500

Qatar 610

Norway 1.200

Mozambique 560

India 2.100

Iceland 810

China 23.300

Germany 500

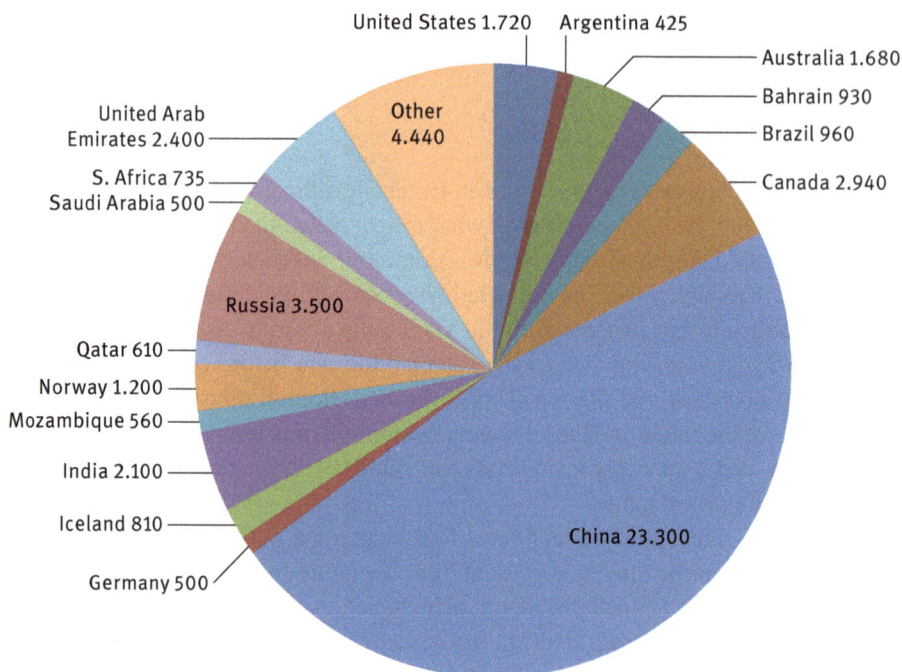

Fig. 11.1: Aluminum production, in thousands of metric tons.

11.2 Refining and isolation, the Hall–Heroult process

More than one aluminum-bearing ore exists, but bauxite is the single source for almost all aluminum metal. Curiously, bauxite often occurs as a red mineral, even though the aluminum in it imparts no color. Rather, iron oxides impart the red color. There are usually other oxides in raw bauxite as well, especially silicates. These are refined out during the early stages of the Hall–Heroult process, and what is called 'red mud' must either be disposed of, or further refined to some economically profitable material, usually iron.

The chemistry of aluminum refining can be presented in a very simple manner, as shown in Scheme 11.1.

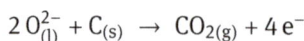

$$2\,O^{2-}_{(l)} + C_{(s)} \;\rightarrow\; CO_{2(g)} + 4\,e^-$$
$$Al^{3+}_{(l)} + 3\,e^- \;\rightarrow\; Al_{(l)}$$
for an overall reaction of
$$6\,O^{2-}_{(l)} + 4\,Al^{3+}_{(l)} + 3\,C_{(s)} \;\rightarrow\; 4\,Al_{(l)} + 3\,CO_{2(g)}$$

Scheme 11.1: Aluminum production, reaction chemistry.

The reality of aluminum refining is significantly more complex. Initially it involves the separation of iron from the bauxite ore using caustic soda (NaOH) at elevated temperature and pressure, which isolates alumina (Al_2O_3). This production of alumina is still called the Bayer process. Next, smelting of the alumina involves a mixture of different fluorides in carbon-lined cells – because the reaction must be run in molten cryolite (Na_3AlF_6) – and involves the oxidation of the carbon electrodes as the aluminum metal is reduced. The cryolite is not generally shown in any reaction chemistry involving aluminum reduction, yet it is essential, as there is no other material or method for aluminum reduction that is as economically feasible. Additionally, molten aluminum must be siphoned out of the cells, since it is denser than the molten cryolite in which the reaction is run [2].

A further factor involved in aluminum reduction from alumina is the electrical requirement. The ArcticEcon website states: "For the Hall–Hérault process to function, an electric current of low voltage but from 200,000 to 500,000 A must pass continuously through each cell. On average it takes about 15.7 kWh of electricity to produce 1 kg of aluminium. This is what makes aluminium smelting such an energy intensive process." (ArcticEcon, https://arcticecon.wordpress.com/) Thus, it is often profitable to locate aluminum refineries close to some source of abundant, inexpensive energy.

At the point at which aluminum metal is formed, the metal can be cast into ingots for future use, or can be immediately alloyed with other metals to make a large variety of different lightweight alloys. Thus, aluminum production has been co-produced on site with metals such as magnesium.

11.3 Uses

Perhaps the most obvious use of aluminum to the consumer is beverage cans. But this is only one of thousands of applications for aluminum. Alcoa lists the following at their website as major uses of aluminum (http://www.alcoa.com/global/en/home. asp):

- Metallic pigments
  - Automotive coatings
  - Electronic materials
  - Packaging – such as aluminum cans
- Chemicals
  - Aluminum powder is used in the synthesis of alcohols and polyolefins
- Rocket propellants
  - Reusable solid rocket
- Photovoltaic thick film pastes
  - For production of solar cells
- Metallurgy
  - Reducing agent – aluminum continues to be used in Goldschmidt reactions

- Heat source
- Alloy component
- Refractories
  - Used in the steel industry, in furnace linings
- Adhesives and sealants
  - Utilized as filler or additive in a variety of adhesives
  - Powders can be used for dry powder coatings
- Explosives
  - Aluminum powder is used in mining and military explosives

This final use of aluminum, in explosives, seems odd to many people, yet the combination of aluminum powder and iron (III) oxide powder is used in what is called the thermite reaction. Scheme 11.2 shows the reaction chemistry.

$$Fe_2O_{3(s)} + 2\,Al_{(s)} \;\rightarrow\; Al_2O_{3(s)} + 2\,Fe_{(l)}$$

Scheme 11.2: The thermite reaction.

Note that the final physical states of the products in Scheme 11.2 include iron as a liquid. When not contained, the thermite reaction is a combustion that produces liquid iron. This has been used to weld the ends of railroad rails together, by initiating the reaction in a metal form clamped to the ends of the rail to be welded, allowing the molten metal to cool, then sanding and grinding the final surface to a smooth finish. When used in what armies call thermite grenades, this becomes an incendiary device that does not stop burning until the reaction is complete.

Another use considered by most to be one of the major uses of aluminum is that of lightweight alloys. Aluminum can be added to another metal to become a minor component in the alloy, or can be the major component. This latter case defines a material as being an aluminum alloy. The number of alloys that have been made and used internationally is large enough that a governing body exists to standardize them [21]. Alloys are given a four digit code. The first digit indicates the major element, with the other digits indicating alloying elements. Thus the 1000 series represents nearly pure aluminum. The 2000 series, for example, are alloyed with copper. The alloy 6061 is a commonly used one, and includes silicon and magnesium, and is 95.85 % aluminum.

11.4 Possible substitutes

Numerous materials, including nonmetallic composites, as well as other alloys, can substitute for aluminum in various applications. In almost all cases, the determining factors for any substitution are whether or not the new material is as light and as strong as aluminum. As in many other cases though, the cost of the substitute material is also factored in to any decision on which material to use.

11.5 Recycling

Aluminum is recycled in almost all developed countries. Consumers tend to think of aluminum beverage cans as the aluminum object most often recycled, but aluminum from windows, doorframes, and other construction applications, as well as aluminum from street lights, tend to weigh more, and thus is almost always recycled. In all cases, aluminum recycling is economically driven, since recycled aluminum costs far less to convert into some end-user object when compared to the refining and reduction of virgin aluminum.

Additionally, aluminum alloy objects are often recycled, usually through scrap yards. In such cases, knowing the composition of the alloy is important, so that it can be again used after any melting and reforming.

Bibliography

[1]   UC Rusal. Website. (Accessed 20 April 2015, as: www.rusal.ru/en).
[2]   Alcoa, Inc. Website. (Accessed 20 April 2015, as: www.alcoa.com).
[3]   Aluminum Corp. of China. Website. (Accessed 20 April 2015, as: www.chalco.com.cn).
[4]   China Power Investment Corp. Website. (Accessed 20 April 2015, as: http://eng.cpicorp.cn).
[5]   Rio Tinto Alcan Inc. Website. (Accessed 20 April 2015, as: www.riotinto.com).
[6]   Norsk Hydro. Website. (Accessed 20 April 2015, as: www.hydro.com).
[7]   China Hongqiao Group Ltd. Website. (Accessed 20 April 2015, as: www.hongqiaochina.com).
[8]   Shangdong Weiqiao Aluminum & Power Co. Website. (Accessed 20 April 2015, as: wq-lygs.alu.cn).
[9]   Shangdong Xinfa Aluminum & Electricity Group Ltd. Website. (Accessed 20 April 2015, as: www.xinfa.cnal.com).
[10] Dubai Aluminum Co. Website. (Accessed 20 April 2015, as: www.dubal.ae).
[11] USGS Mineral Commodity Summaries. Downloaded as: http://minerals.usgs.gov/minerals/pubs/mcs/2015/mcs2015.pdf
[12] World Aluminium. Website. (Accessed 8 September 2015, as: http://www.world-aluminium.org/statistics/).
[13] The Aluminum Association. Website. (Accessed 8 September 2015, as: http://www.aluminum.org).
[14] European Aluminum Association. Website. (Accessed 8 September 2015, as: http://www.alueurope.eu).
[15] Australian Aluminum Council. Website. (Accessed 8 September 2015, as: http://aluminium.org.au).
[16] Aluminum Association of Canada, Dialog on Aluminum. Website. (Accessed 8 September 2015, as: http://www.thealuminiumdialog.com/en/about-the-aac/our-mission).
[17] Aluminum Federation of South Africa. Website. (Accessed 8 September 2015, as: http://www.afsa.org.za/).

[18] Aluminium Association of India. Website. (Accessed 8 September 2015, as: http://www.aluminium-india.org/).
[19] Alcoa, education. Website. (Accessed 25 September 2015, as: http://www.alcoa.com/global/en/about_alcoa/pdf/startswithdirt.pdf).
[20] ArcticEcon. Website. (Accessed 25 September 2015, as: https://arcticecon.wordpress.com/2012/02/15/aluminium-smelting-in-iceland-alcoa-rio-tinto-alcan-century-aluminum-corp/).
[21] Aluminum.org. "International Alloy Designations and Chemical Composition Limits for Wrought Aluminum and Wrought Aluminum Alloys," Website. (Accessed 25 September 2015, as: http://www.aluminum.org/sites/default/files/TEAL_1_OL_2015.pdf).

12 Titanium

12.1 Introduction

Titanium has less history than many elemental metals, only having been reported for the first time in 1791. But after its discovery, this new metal rapidly found uses in a variety of industrial applications because of its high strength, hardness, and resistance to corrosion even in extreme environments. Additionally, the metal has a relatively low density, 4.506 g/cm^3, which makes it useful as an elemental material, or in alloys with other low density metals in applications where overall weight of the end item is an important variable.

Titanium is one of only a few metals that are used as extensively or more as a compound than as a metal. In the case of titanium, the compound is TiO_2. We will discuss this along with the metal, because of the large-scale uses of it.

12.2 Location and sources

Small amounts of titanium occur in several different minerals, but the two from which it is profitably extracted are rutile, TiO_2, and ilmenite, $FeTiO_3$, although the formula for ilmenite can also be expressed as $(Fe, Mg, Mn, Ti)O_3$, depending on the specific source. Within the United States there are several operations that produce titanium or what is called titanium sponge. Figure 12.1 shows the world production of titanium,

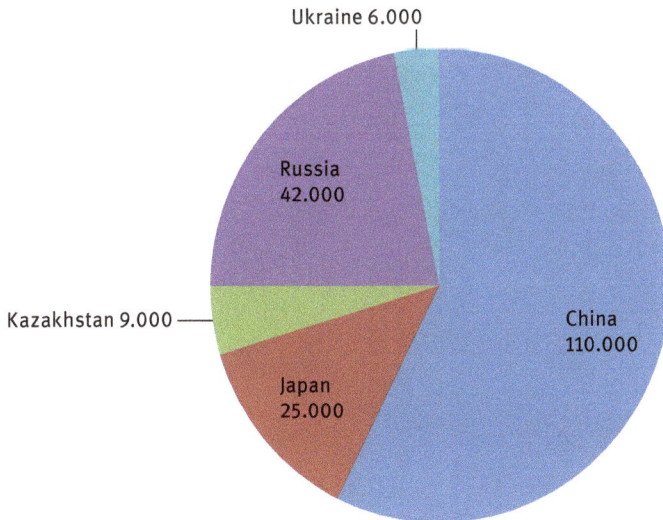

Fig. 12.1: Titanium production worldwide, in metric tons.

but excludes that of the USA, because firms in the United States did not disclose their production statistics, citing proprietary concerns [1].

A list of the major ilmenite producers worldwide includes the following:

1. Tellnes Mine, Sokndal, Norway
2. Lac Tinto Mine, in the Rio Tinto Group, Quebec, Canada
3. Richards Bay Minerals, KwaZulu-Natal, South Africa
4. Moma mine, Kenmare Resources, Mozambique
5. Murray Basin, and Eneabba, of Iluka Resources, Australia
6. Indian Rare Earth Mineral Mines, Kerala, India
7. Grande Cote Mine, of TiZir, Ltd., Senegal
8. QIT Madagascar Minerals Mine, in the Rio Tinto Group [2–9]

As mentioned, there are other deposits as well, often somewhat smaller. Those listed above are both mining operations and those that produce ilmenite from sands.

Ilmenite has also been found to exist on the surface of our Moon, especially in some of the seas, the maria. Its reduction to titanium has been proposed, although such proposals usually are accompanied by the idea of capturing the oxygen in the

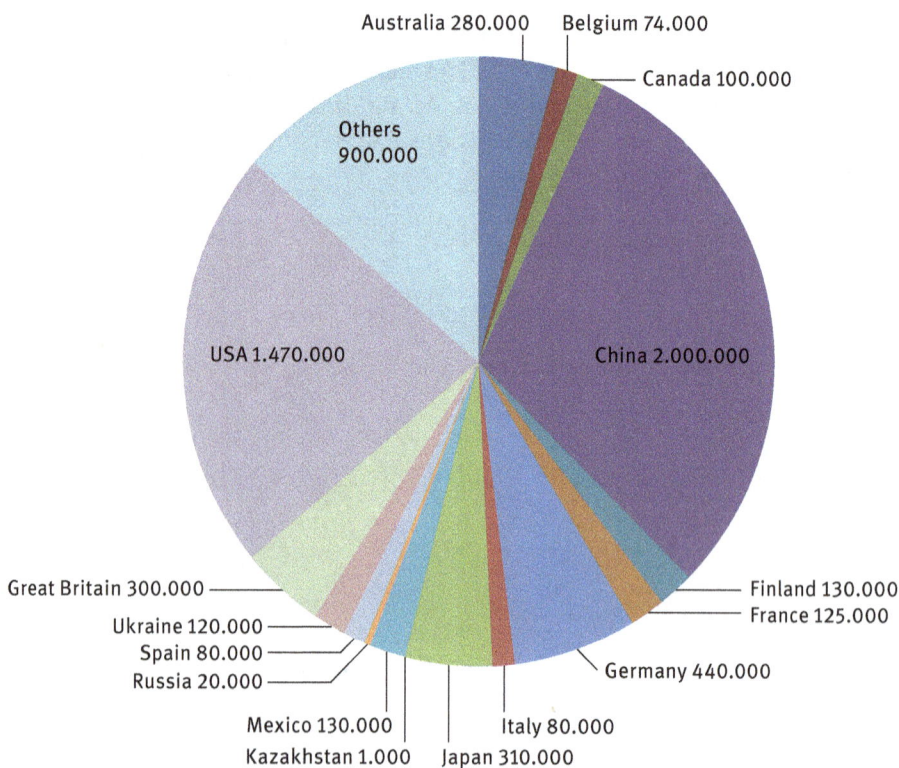

Fig. 12.2: Titanium dioxide production.

ore as a means of producing oxygen for use by human colonists or explorers on the Moon. Perhaps obviously, no corporation has yet progressed in its planning to the point at which this can be attempted.

Titanium is also refined directly into titania (TiO_2), sometimes called titanium dioxide or titanium pigment, as well as into titanium metal. While this formula is the same as that for rutile ore, remember that rutile does not occur as pure TiO_2, since it includes traces of other elements in varying amounts, depending on the ore batch (as do almost all other ores). The US Geologic Survey tracks this as well, because there are numerous uses for this oxide material [1]. Figure 12.2 shows this for worldwide production, again in terms of metric tons of the material.

12.3 Refining and isolation

Titanium refining can be effected through two different processes. The Kroll process is by far the most economically advantageous method, although we will discuss the older Hunter process as well.

12.3.1 The Kroll process

Titanium is widely but unevenly distributed throughout the world, with only a few countries producing the world's supply. They were shown in Figure 12.1.

The Kroll process is the current means by which titanium is reduced from its ores, wherever they are located. The chemistry can be represented in a straightforward manner, as shown in Scheme 12.1. The titanium dioxide starting material has been refined to remove iron impurities prior to the reaction, in what is called the Becher process. The carbon is supplied from coke. As with many metal extraction processes, it is a pyrometallurgical one, meaning it is run at high temperature.

$$TiO_{2(s)} + C \rightarrow Ti_{(s)} + CO_{2(g)} \qquad \text{at } 1,000\,°C$$

which is not isolated, but rather captured as

$$Ti_{(s)} + 2\,Cl_{2(g)} \rightarrow TiCl_{4(g)}$$

Scheme 12.1: The Kroll process.

This product is usually combined with different volatile chlorides that must then be distilled. This distillation can be accomplished with relative ease because the melting point of $TiCl_4$ is $-24\,°C$ and its boiling point is $136\,°C$. After this step, elemental titanium can be isolated as shown in Scheme 12.2, in a stainless steel reaction apparatus.

$$TiCl_{4(g)} + 2\,Mg_{(l)} \;\rightarrow\; 2\,MgCl_{2(l)} + Ti_{(s)} \qquad \text{at } 800\text{--}850\,°C$$

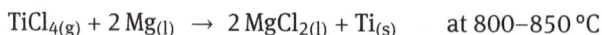

Scheme 12.2: Titanium isolation, the final step of the Kroll process.

Since this final step is run in stainless steel reactors, or molybdenum-lined reactors, which are essentially corrosion resistant, the container also ensures maximum reduction of the titanium. The product is referred to as titanium sponge. This material is crushed to uniform particle size, then brought to a molten state in a vacuum arc furnace, then allowed to cool and form in ingots, again under an inert atmosphere. The inert atmosphere is important, as titanium oxides or nitrides can form should the metal be exposed to air or traces of water while cooling.

12.3.2 The Hunter process

The Kroll process has proven itself to be economically better for the production of titanium than the Hunter process, but this latter, older process produces a highly pure form of the element, and was the first process ever to produce pure titanium. In Matthew Hunter's paper on the subject, published in 1910, the author states: "All the earliest attempts at the preparation of metallic titanium resulted for the most part in the production of the various nitrides which from their metallic appearance were always mistaken for the metal [10]." The reaction chemistry for the process can be shown as follows, in Scheme 12.3.

$$TiCl_{4(l)} + 4\,Na_{(l)} \;\rightarrow\; 4\,NaCl + Ti_{(s)} \qquad \text{at } 700\text{--}800\,°C$$

Scheme 12.3: The Hunter process.

Note that since titanium tetrachloride is a liquid at ambient temperature, it can be isolated and used as well if need be. Numerous uses have been found for this material, $TiCl_4$, often in different organic chemistry reactions.

12.4 Uses

12.4.1 High strength alloys

Titanium is perhaps best known for its uses in high strength, lightweight alloys. So many alloys that incorporate titanium have been produced that there are now standardized tables for them [11–13].

Since many titanium alloys are produced for their low density, and ultimately for their light weight in user end products, titanium alloys can in several cases and classes be considered aluminum alloys. This is because aluminum is another low-density metal that has found a wide variety of uses in many industries and consumer products [11, 12]. Even though the use of these two lightweight metals, either as elemental metals or as alloys, spans a wide range of end items, the general populace tends to think of aircraft and the aerospace industry as one of the major users of such materials. An International Titanium Association exists in part to promote the use of titanium and titanium-based alloys, and at their website devotes a significant amount of space to how titanium is used in the aerospace industry [14].

Additionally, ruthenium has been found to be a metal that improves the high strength properties of titanium when mixed as an alloy. Even titanium alloys with as little as 0.1 % ruthenium have been found to be up to 100 times stronger than pure titanium. This has become a niche use for a metal, ruthenium, an element that really has no other major industrial use.

12.4.2 Pigments, titanium white or titanium dioxide

The pigment known as titanium white represents a major use for titanium in industry, in the form of a titanium compound. Titanium white produces a color that is flat, long-lasting, and color fast. Some of the major companies that mine ilmenite and rutile even advertise themselves as producers of titanium white more so than of titanium metal. For example Kenmare states of its Moma mining operation, on its website:

"The Moma Mine contains globally significant reserves of heavy minerals which include the titanium minerals ilmenite and rutile (primarily used to produce titanium dioxide pigment), as well as the relatively high-value zirconium silicate mineral, zircon. Titanium dioxide pigment is used in paints, paper and plastic production. The primary applications for zircon are in the manufacture of opacifiers for ceramic tile production and for refractory products used in the steel and foundry industries [5]."

Clearly, titanium white pigment is the major use in such an operation.

Both titanium sponge and titanium dioxide are imported to the United States, and the imports are tracked, once again by the USGS Mineral Commodities Summary. Imports of titanium dioxide pigments are as follows: Canada 39 %; China 19 %; Germany 7 %; Others 35 % [1].

12.5 Possible substitutes

The USGS Mineral Commodity Summaries notes that there are few substitutes for titanium in many applications, but states: "… titanium competes with aluminum, composites, intermetallics, steel, and superalloys [1]." This is in respect to the metal. Interestingly, titanium white is so prevalent in many industries and applications that it too is mentioned in the Mineral Commodity Summaries, where it states: "Ground calcium carbonate, precipitated calcium carbonate, kaolin, and talc compete with titanium dioxide as a white pigment [1]."

12.6 Recycling

Titanium metal generally costs five to six times as much as stainless steel, and thus it is almost always recycled. Recycled titanium is used almost exclusively by the steel industry as well as in superalloys [14].

Bibliography

[1]   USGS Mineral Commodity Summaries. Downloaded as: http://minerals.usgs.gov/minerals/ pubs/mcs/2015/mcs2015.pdf
[2]   Kronos. Website. (Accessed 15 October 2015, as: http://kronostio2.com/en/manufacturing-facilities/hauge-norway).
[3]   Rio Tinto, Quebec. Website. (Accessed 15 October 2015, as: http://www.riotinto.com/ diamondsandminerals/rio-tinto-fer-et-titane-14778.aspx).
[4]   Rio Tinto, Richards Bay. Website. (Accessed 15 October 2015, as: http://www.riotinto.com/ diamondsandminerals/richards-bay-minerals-4642.aspx).
[5]   Kenmare. Website. (Accessed 15 October 2015, as: http://www.kenmareresources.com/ operations/summary-of-operations.aspx).
[6]   Iluka. Website. (Accessed 15 Oct 2015, as: https://www.iluka.com/company-overview/ operations/perth-basin-western-australia).
[7]   Indian Rare Earth Minerals. Website. (Accessed 15 October 2015, as: http://www.irel.gov.in/ scripts/unit.asp).
[8]   TiZir, Ltd. Website. (Accessed 15 October 2015, as: http://www.tizir.co.uk/projects-operations/ grande-cote-mineral-sands/).
[9]   Rio Tinto, Madagascar. Website. (Accessed 15 October 2015, as: http://www. riotintomadagascar.com/english/index.asp).
[10] Hunter, M. A.: Metallic Titanium. *J. Am. Chem. Soc.*, 1910, 32(3), pp 330–336.
[11] Alcoa. Website. (Accessed 13 April 2015, as: http://www.alcoa.com/global/en/home.asp).
[12] Alcoa, titanium. Website. (Accessed 8 October 2015, as: http://www.rtiintl.com/Titanium/RTI-Titanium-Alloy-Guide.pdf).
[13] ASTM International. Website. (Accessed 8 October 2015, as: http://www.asminternational.org/ documents/10192/1849770/ACFA9E7.pdf).
[14] International Titanium Association. Website. (Accessed 14 October 2015, as: http://www. titanium.org/).

13 Magnesium

13.1 Introduction

Like many metals, magnesium does not have a particularly long history, having first been discovered in 1808 by Sir Humphry Davy and isolated after tedious work (although hydrated magnesium sulfate – Epsom salt – had been known for almost two hundred years prior to this). But magnesium has become extremely important in a variety of metal alloys. Since it is a rather abundant element in the Earth's crust, it can be mined in one mineral form or another in many locations.

13.2 Refining and isolation

Magnesium metal is often refined from dolomite (which is $CaMg(CO_3)_2$) and magnesite (which is $MgCO_3$), and exists in more than fifty other minerals as well. Not all magnesium-containing minerals are economically profitable to mine and refine to the metal, however. A non-exhaustive list of those that have been mined for magnesium in the past is shown in Table 13.1.

As well, magnesium in ionic form is present in seawater, in lesser amounts than sodium, but still in large enough quantities that seawater too can be a commercially viable source of the metal, depending on other costs, such as that for electricity.

Magnesium production worldwide is shown in Figure 13.1, in thousands of metric tons. The U.S. magnesium plant in Utah is not listed in this figure, as the company cites proprietary concerns in reporting its data.

It is obvious that China currently dominates the world market for this metal.

The reduction of magnesium ores from different minerals is sometimes called the Pidgeon process, and involves either reduction using ferrosilicon alloys, or reduction using carbon. The reaction chemistry can be presented rather simply, although there

Table 13.1: Major magnesium-containing minerals [1, 2].

Name	Formula	Location
Brucite	$Mg(OH)_2$	Pennsylvania, USA
Carnallite	$KMgCl_3 \cdot 6(H_2O)$	USA, Germany, Russia, Canada
Dolomite	$CaMg(CO_3)_2$	China, Australia, Congo, Morocco, USA, Canada, Brazil, Mexico
Magnesite	$MgCO_3$	South Australia, Brazil, China, India, Russia, Turkey, Spain, Greece, Slovakia, Austria
Olivine	$(Mg, Fe)_2SiO_4$	USA, China, Brazil, Australia, Kenya, Egypt, Mexico, Norway, South Africa, Pakistan, Sir Lanka, Tanzania
Talc	$Mg_3Si_4O_{10}(OH)_2$	Australia, France, Brazil, Oman, Turkey

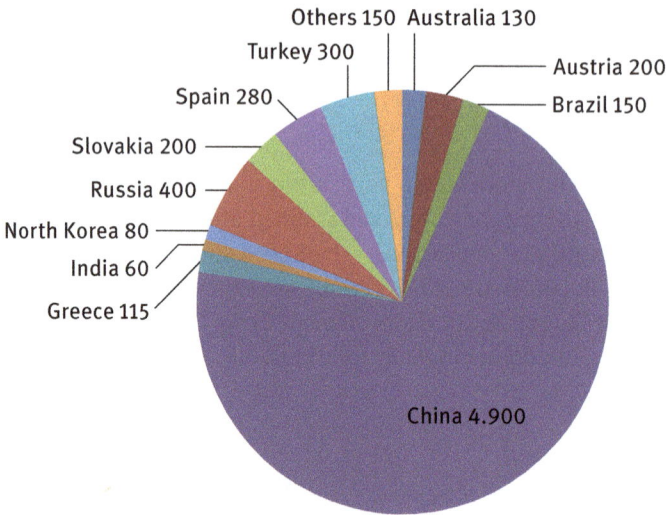

Fig. 13.1: Magnesium production worldwide, in thousands of metric tons.

are details beyond the reactions that cannot be ignored. Scheme 13.1 shows the simplified reactions for magnesium reduction and refinement.

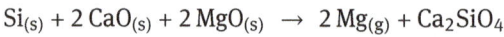

$$Si_{(s)} + 2\,CaO_{(s)} + 2\,MgO_{(s)} \ \rightarrow \ 2\,Mg_{(g)} + Ca_2SiO_4$$

or

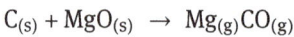

$$C_{(s)} + MgO_{(s)} \ \rightarrow \ Mg_{(g)}CO_{(g)}$$

Scheme 13.1: Production of magnesium metal.

The first reaction is usually run at approximately 1,200–1,500 °C in stainless steel retorts. The iron in the ferrosilicon alloy does not participate in the reaction other than to deliver silicon to it. The gaseous magnesium must then be condensed and solidified. When carbon is used, the reaction temperature is usually higher, around 2,300 °C. Once again, the gaseous magnesium is cooled and then formed into ingots as needed.

As mentioned, magnesium can also be extracted from seawater. Scheme 13.2 shows the simplified reaction chemistry for this. This process is known as the Dow process, and represents an electrolytic reduction.

$$Ca(OH)_2 + MgCl_2 \ \rightarrow \ Mg(OH)_{2(s)} + CaCl_{2(aq)}$$

followed by

$$Mg(OH)_{2(s)} + 2\,HCl \ \rightarrow \ 2\,H_2O + MgCl_{2(aq)}$$

followed by

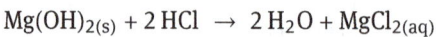

$$MgCl_2 \ \rightarrow \ Mg + Cl_{2(g)}$$

Scheme 13.2: Production of magnesium from seawater.

The first two reactions appear to be a cycle, producing the first reactant. But the starting magnesium chloride is co-mingled with the much more common sodium chloride. Magnesium is usually present in seawater to only about 12 % that of sodium. Thus, separating it by solubility, to insoluble magnesium hydroxide, then returning it to an aqueous state allows the electrolysis – the final reaction – to start with a pure magnesium compound, and result in reduced magnesium metal.

Additionally, solid oxide membrane (SOM) technology has recently been proven to work for the reduction of magnesium directly from magnesium oxide. The co-product is elemental oxygen, the result of the electrolysis being performed using a yttria-stabilized zirconia electrolyte. This reaction runs at slightly lower temperature than the established methods, and should reduce the cost of production. It has not yet been scaled up by major magnesium producers.

13.3 Uses

While there are many uses for magnesium in compounds – with perhaps the most obvious one for the general public being either talc as a body powder, or perhaps Epsom salts – we will confine the discussion to the uses of magnesium as an element or in alloys. Concerning magnesium as a metal, the USGS Mineral Commodity Summaries breaks down uses to:
1. "…aluminum-based alloys that were used for packaging, transportation, and other applications…," 35 %
2. "…as a reducing agent for the production of titanium and other metals…," 30 %
3. "Structural uses…(castings and wrought products)…," 15 %
4. "…desulfurization of iron and steel…," 10 %

With all other uses comprising the remaining 10 % [1].

13.3.1 Elemental magnesium

Magnesium metal is used as a major part of the Kroll process for titanium reduction and refining, as seen in Chapter 12. It can also be used for the reduction of other metals, acting as the reducing agent, because it is easily oxidized, and because its chloride can be re-reduced to the metal for repeated uses in such processes.

Because of the intensely bright, white light that magnesium produces when it burns, it has found uses in traditional flash photography, emergency flares, and fireworks, although this is no longer a major use of it.

13.3.2 Magnesium–aluminum alloys

Magnesium has been alloyed with aluminum and other metals, largely in the search for lightweight metal alloys. Since its density is a very low $1.738\,g/cm^3$, it has been used in aircraft manufacture from the days of the earliest metal airplanes. For the purposes of comparison, aluminum has a density of $2.70\,g/cm^3$, iron has $7.874\,g/cm^3$, and gold has $19.30\,g/cm^3$. Table 13.2 lists a number of magnesium alloys, and the properties they possess.

Table 13.2: Magnesium alloy properties [3].

Alloyed with:	Advantages	Disadvantages	Comments
Aluminum	Strength and hardness	Increasing Al = decreasing ductility	Most common alloy class
Calcium	High temp. performance	Increasing Ca = decreasing malleability	
Cerium	Corrosion resistance	Lower strength	
Copper	Strength	Lower ductility	
Manganese	Corrosion resistance		
Nickel	Strength	Ductility and corrosion resistance	
Neodymium	Strength	Cost	
Silicon	High temp. performance		
Tin	Ductility, malleability		
Yttrium	High temp strength		
Zinc	Corrosion resistance		Second most common alloy class

13.3.3 Castings

Magnesium parts or magnesium alloy parts are often cast. Indeed, some magnesium alloys are used precisely because they have superior flow when forming end items than other metals or alloys. Often, the molds are segmented into two or more pieces so that they may be reused many times, the metal is cast into them, the molds are removed when the metal has solidified, then further trimming or shaving of the item must take place at the seam where the molds met. Some final polishing may be required as well.

The International Magnesium Association lists the following as major uses of the metal:

- Aircraft and missile components
- Aircraft engine mounts, control hinges, fuel tanks, wings

- Automotive wheels, housings, transmission cases, engine blocks
- Bicycles and other sporting goods equipment
- Equipment for material handling
- Ladders
- Laptops, televisions, cell phones
- Luggage
- Portable power tools, chainsaws, hedge clippers, weed whackers
- Printing and textile machinery
- Steering wheels and columns, seat frames [3]

While this may not seem to be in direct alignment with the statement made by the USGS, it should be noted that the two organizations track magnesium differently, one in terms of government need, and the other in terms of economics and sale of end-user items.

13.3.4 Automotive

Magnesium has found use in automotive parts, in all cases where both lower weight and high strength are important. However, the classic 'mag wheels' of the past are now essentially aluminum wheels with small amounts of magnesium incorporated, making an alloy.

13.3.5 Aerospace

As with automotive applications, magnesium is used in the aerospace industry when high strength and low density of materials is of paramount concern. Because of its reactivity however, much of the lightweight metal material in aircraft can be classified as aluminum alloys with magnesium in them.

13.3.6 Electronic

Curiously, magnesium does find use in modern, small electronic devices, but not usually as a component in the electronic circuitry. Rather, because magnesium is high strength and low density (which for the size of the device usually translates to low weight), it has become more prevalent as the case or housing of devices such as cell phones in the past five years. Aluminum–magnesium alloys can also be used, because both do an excellent job of protecting devices from the accidentally abusive treatment they too often receive.

13.4 Possible substitutes

We have seen that magnesium is usually refined because of its low density, and thus light weight in all applications. Thus, aluminum and titanium compete with it in terms of uses of the metal in end items that are cast, as well as wrought metal items.

13.5 Recycling

Magnesium in the form of alloyed metal is almost always recycled. As with most recycling, the driving force is an economic one. Scrapyards that deal with metals and alloys, especially those dealing with steel and aluminum, generally deal with magnesium alloys as well. The USGS Mineral Commodity Summaries states: "In 2014, about 25,000 tons of secondary magnesium was recovered from old scrap and 57,000 tons were recovered from new scrap [1]."

Additionally, magnesium that has been oxidized to magnesium chloride during the production of titanium is again reduced to magnesium metal, simply so that it can again be used in titanium refining.

Bibliography

[1] USGS Mineral Commodity Summaries. Downloaded as: http://minerals.usgs.gov/minerals/pubs/mcs/2015/mcs2015.pdf
[2] Australian Mine Atlas. Website. (Accessed 23 October 2015, as: http://www.australianminesatlas.gov.au/education/fact_sheets/magnesium.html).
[3] International Magnesium Association. Website. (Accessed 16 April 2015, as: http://www.intlmag.org/).
[4] International Magnesium Association, Alloys. Website. (Accessed 15 October 2015, as: http://www.intlmag.org/magnesiumresources/alloys.cfm).

14 Uranium and thorium

14.1 Introduction

The elements thorium and uranium were the two actinides discovered first, and indeed were discovered before almost all of the lanthanides as well. Uranium was first reported in 1789 and thorium in 1828. The most stable isotope of thorium has a half-life of more than 14 billion years, and uranium-238 has a half-life of longer than 4.4 billion years, which means that based on the age of the planet, they are quite stable, could be extracted from different ores as they were mined, and had not decayed to any appreciable extent by the time they were isolated and determined to be new elements. Importantly, both of these elements occur as major components of several ores, while the lanthanides are more evenly dispersed in small amounts in a very wide variety of ores.

Despite their long half-lives, both uranium and thorium are radioactive, and this phenomenon was one of the driving forces for early experimentation with them. With uranium, the material form was potassium uranyl sulfate. Henri Becquerel was the first to deduce correctly that radiation was coming from a uranium source. Marie Curie appears to be the first to use the term 'radioactivity', and Ernest Rutherford the first to categorize radioactive emissions in terms of their ability to penetrate matter, as alpha, beta, and gamma radiation.

14.2 Refining and isolation

Both elements require several steps for their refining and purification. Here, each is treated separately because of the differences in the processes.

14.2.1 Uranium isolation and refining

Large uranium mines exist in several different countries. The top ten mines worldwide are listed in Table 14.1. While the USGS Mineral Commodity Summaries does not track uranium production, the World Nuclear Association does [1–5].

It should also be noted that while uranium production is tracked, specific mineral and ore types usually are not. There are several different minerals that contain uranium. A non-exhaustive list is shown in Table 14.2. Pitchblende is the mineral containing the highest percentages of uranium, and throughout the world, the deposits in Canada have been proven to be the richest. The percent of uranium in each mineral is listed in Table 14.2, but it should be remembered that ores are not always pure, and thus the actual percentage may be much lower than listed, depending on the ore

Table 14.1: Top ten uranium mines worldwide.

No.	Name	Location	Amt. (tons)	Other products
1	McArthur River, Areva-Cameco	Canada	7,654	
2	Ranger, Energy Resources of Australia	Australia	3,216	
3	Rossing, Rio Tinto	Namibia	3,077	
4	Kraznokamensk	Russia	2,920	
5	Arlit	Niger	3,000 est.	
6	Tortkuduk, Kazatomprom-Areva	Kazakhstan	2,439	
7	Olympic Dam, BHP Billiton	Australia	2,330	Copper, gold, silver
8	Budenovskoye 2, Kazatomprom	Kazakhstan	1,708	
9	South Inkai, Uranium One	Kazakhstan	1,701	
10	Inkai, Kazatomprom-Cameco	Kazakhstan	1,642	

Table 14.2: Uranium-bearing minerals.

Name	Formula*	% Uranium	Location
Autunite	$Ca(UO_2)_2(PO_4)_2 \cdot 8\,H_2O$	52.1	France, USA
Brannerite	UTi_2O_6	55.3	
Carnotite	$K_2(UO_2)_2(VO_4)_2 \cdot 3\,H_2O$	52.8	USA, Congo, Morocco, Australia, Kazakhstan
Coffinite	$USiO_4(OH)_4$	80.4	USA
Davidite	$(Ln)(Y, U)(Ti, Fe^{3+})_{20}O_{38}$ **		Australia, Norway
Saleeite	$Mg(UO_2)_2(PO_4)_2 \cdot 10\,H_2O$	51.0	Congo
Thucholite	Mix of uraninite, hydrocarbons and sulfides		Canada
Torbernite	$Cu(UO_2)_2(PO_4)_2 \cdot 12\,H_2O$	47.1	France, USA
Tyuyamunite	$Ca(UO_2)_2(VO_4)_2 \cdot 8\,H_2O$	50.0	Kyrgystan (rare)
Uraninite (aka pitchblende)	UO_2	88.1	Congo, Canada, Australia
Uranocircite	$Ba(UO_2)_2(PO_4)_2 \cdot 10\,H_2O$	45.5	Germany
Uranophane	$Ca(UO_2)_2(HSiO_4)_2 \cdot 5\,H_2O$	55.6	
Zeunerite	$Cu(UO_2)_2(AsO_4)_2 \cdot 10\,H_2O$	44.8	Germany

* formulas are given in their simplest form,   ** Ln = some lanthanide.

batch. Concentrating an ore prior to the reduction chemistry is always part of the process in isolating this metal. As well, it can be seen from Table 14.2 that for some of the minerals, other economically worthwhile elements, such as the rare earths, must be separated during the refining process.

Like many metals, the refining of uranium involves the separation of the element from other components in its ores, followed by chemical reduction to the metal. Broadly, one can say it is separated as UO_2, then reduced to the element. In somewhat simplified form, the steps are shown in Schemes 14.1 and 14.2 , with 14.1 stopping at the production of what is called yellow cake.

$$UO_3 + 3\,H_2SO_4 + 3\,H_2O \;\rightarrow\; H_4[UO_2(SO_4)_3] \qquad \text{acid treatment}$$

or

$$UO_3 + 2\,NaHCO_3 + Na_2CO_3 \;\rightarrow\; Na_4[UO_2(CO_3)_3] \qquad \text{caustic treatment}$$

followed by

$$2[UO_2(SO_4)_3]^{4-} + 3\,H_2O + 6\,NH_3 \;\rightarrow\; (NH_4)_2U_2O_7 + 2(NH_4)_2SO_4 + 4\,SO_4^{2-}$$

or

$$2[UO_2(CO_3)_3]^{4-} + 14\,NaOH \;\rightarrow\; Na_2U_2O_7 + 3\,H_2O + 6\,Na_2CO_3 + 8\,OH^-$$

Scheme 14.1: Uranium isolation and reduction.

The complex formed in the first step of Scheme 14.1 must be passed through an anion exchange resin to ensure its purity. The product at the end of either reaction in the second step is a precipitate, which when dried and collected is known as yellow cake.

$$(NH_4)_2U_2O_7 + 4\,HNO_3 \;\rightarrow\; 2\,UO_2(NO_3)_2 + 2\,NH_3 + 3\,H_2O$$

then extraction with tributylphosphate

$$2(H_9C_4)_3PO_4 + UO_2(NO_3)_2 \;\rightarrow\; UO_2(NO_3)_2((H_9C_4)_3PO_4)_2$$

then, evaporation to $UO_2(NO_3)_2$

then

$$UO_2(NO_3)_2 \;\rightarrow\; UO_3 \qquad \text{at } 300\,°C$$

then, reduction with hydrogen

$$UO_3 + H_2 \;\rightarrow\; UO_2 + H_2O \qquad \text{at } 700\,°C$$

then reacting with anhydrous HF

$$UO_2 + 4\,HF \;\rightarrow\; UF_4 + 2\,H_2O \qquad \text{at } 550\,°C$$

and finally, reduction with magnesium

$$UF_4 + 2\,Mg \;\rightarrow\; 2\,MgF_2 + U \qquad \text{at } 700\,°C$$

Scheme 14.2: Uranium reduction from yellow cake.

The first steps in Scheme 14.2 isolate pure uranyl nitrate, then after heating to produce uranium trioxide, the subsequent steps are the reduction to the metal. The final step is chemically much like the Kroll process for titanium production.

14.2.2 Thorium isolation and refining

Thorium is found in several different areas of the world but is seldom mined as the primary material in an ore. Figure 14.1 shows the current world reserves of thorium,

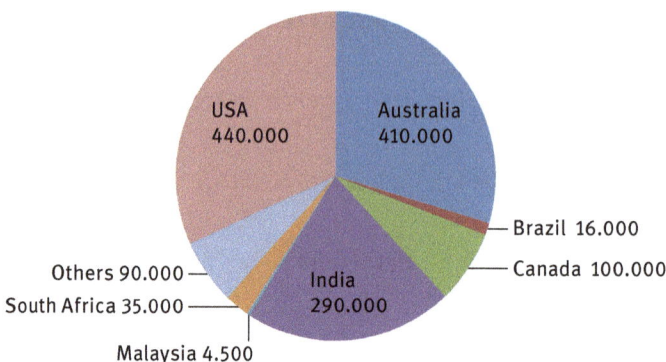

Fig. 14.1: Thorium reserves, in metric tons.

in metric tons. Note that this is not thorium refined for specific uses, but rather reserves that can later be developed. Since thorium has not had any large-scale use in the past, these numbers remain somewhat speculative, since no corporation has actively sought out thorium ores worldwide.

Thorium occurs in numerous ores that are usually mined for their rare earth elements (lanthanides), which will be discussed in Chapter 15. This co-location of the elements exists because thorium has several chemical similarities to the lanthanides, even though it is an actinide. Despite its wide occurrence in numerous ores, thorium today is generally mined from monazite sands, some of which can contain nearly 20 % ThO_2, usually called thoria.

In general, the processing steps for thorium are difficult to represent as stoichiometric reactions, because they separate out other elements, usually based on differences in solubility in acidic or basic conditions. Broadly, thorium is refined as follows:
1. Monazite is dissolved in H_2SO_4 at elevated temperature.
2. The solution is dissolved in water at 0 °C.
3. NH_3 is then added, precipitating thorium salts.
4. Thorium-containing salts are dissolved in HNO_3.
5. Thorium is extracted from the acidic solution as $[Th(NO_3)_4 \cdot PO_4(C_4H_9)_3]$ by using tributylphosphate.
6. The complex is then heated to produce thorium oxide, ThO_2.
7. Under argon at 1,000 °C calcium is reacted with ThO_2 to produce the metal. This can be shown as: $2\,Ca + ThO_2 \rightarrow Th + 2\,CaO$.

Once again, the final step is much like the final step of the Kroll process, with a metal serving as the reducing agent.

14.3 Uses

14.3.1 Power generation

Uranium is used almost exclusively for electric power generation and as the basis of atomic weapons. Power generation in a nuclear plant works on the principle of using isotopically enriched uranium, which then produces heat in water in a closed loop system. When the water boils, the steam is used to turn a turbine's blades. From this point, the production of power is based on the turbines.

Thorium has not yet been used as a major source of nuclear power, but it has the potential to be such a source. The reason enriched uranium has been the most completely developed source of nuclear power has its origin in the Manhattan Project of World War II, and the United States' effort to develop an atomic weapon. In the course of the weapon's development, it was realized that slightly lower enrichment would be sufficient to produce the steam needed to generate electric power.

14.3.2 Weaponry

Uranium must be enriched to a higher isotopic purity of uranium-235 than is necessary for the production of electricity in order to produce what is called weapons-grade uranium. High enough enrichment enables the uranium to produce power without the limitations required for electricity production, and the resulting energy release is an atomic explosion.

Additionally, what are called depleted uranium rounds (DU) are conventional projectile tank shells that have been made from the non-fissile uranium-238 that is separated from the fissile uranium-235. The United States and some NATO militaries adopted this type of ammunition after it was found to be superior at penetrating the armor of enemy tanks when compared to other metals. This is because of uranium's high density. The ability to produce such ammunition is simply because a great deal of uranium-238 already exists, and there were no other uses for it. These rounds were first fielded in the Gulf War of 1991.

14.3.3 Compounds and uses

Prior to World War II, uranium had been used as a glass additive for decorative glass objects. Depending on its concentration in the glass matrix, it produces yellows or yellow-green colors.

Thorium has several smaller scale uses. They include the following:
- Lantern mantles. As thoria, this application produces intense, white light. These mantles are often called Welsbach mantles.

- Ceramics. Thoria is used, as opposed to the reduced metal, and results in ceramics that are highly heat resistant.
- Alloy. Thorium can be used as an alloying agent to create metal welding rods that burn at specific temperatures.
- Mag-thor alloy. Thorium can also be alloyed with magnesium and zirconium to produce high strength alloys at high temperature. Such alloys find use in the aerospace industry.
- Coating. Thorium is at times used to coat or plate tungsten filaments. Such devices find use in the electronics industry.
- Crucibles. Thoria is used because of its high melting point, around 3,300 °C.
- Camera lenses. Some high quality camera lenses utilize thoria in their make-up.

While each of these applications requires only a small amount of thorium in the end product, and while the radioactivity level is safe for such small amounts, this is not true when many of a particular item are gathered and concentrated. The book, *The Radioactive Boy Scout*, tells the story of David Hahn, a high school student who at age 17 tried to make a breeder reactor in his home by purchasing large numbers of thorium mantles and concentrating the thorium. The book also discusses the resulting radioactive pollution that resulted from his efforts [6].

14.4 Possible substitutes

In terms of producing energy, there is currently no substitute for enriched uranium, although India may soon bring a thorium-based nuclear power plant on line.

Thorium can be replaced by other metals in alloys, as long as the final product has the same desired properties as the thorium-containing alloy. Yttrium has been used successfully in some cases.

The depleted uranium ammunition that was mentioned can be substituted by other, high density, armor-piercing tank rounds.

14.5 Recycling

Thorium and thorium-containing compounds are not currently recycled by any country or state.

Uranium and uranium-containing materials have not traditionally been recycled, with the exception of uranium fuel rods. At its website, the World Nuclear Association discusses what is called the re-processing of nuclear fuel. It states:

"A key, nearly unique, characteristic of nuclear energy is that used fuel may be reprocessed to recover fissile and fertile materials in order to provide fresh fuel for existing and future nuclear power plants. Several European countries, Russia and Japan have had a policy to reprocess used nuclear fuel, although government policies in many other countries have not yet come round to seeing used fuel as a resource rather than a waste [3]."

The recycling of uranium fuel rods is once again often a matter of economics. If it is more costly to produce new metal, re-processing spent fuel rods can become a viable option.

Bibliography

[1] USGS Mineral Commodity Summaries. Downloaded as: http://minerals.usgs.gov/minerals/ pubs/mcs/2015/mcs2015.pdf
[2] International Thorium Energy Organization. Website. (Accessed 16 April 2015, as: http://www. itheo.org/).
[3] World Nuclear Association. Website. (Accessed 16 April 2015, as: http://www.world-nuclear. org/).
[4] American Nuclear Society. Website. (Accessed 30 October 2015, as: http://www.ans.org/).
[5] United States Nuclear Regulatory Commission. Website. (Accessed 16 April 2015, as: http:// www.nrc.gov/).
[6] Silverstein K. The Radioactive Boy Scout: The True Story of a Boy and His Backyard Nuclear Reactor, 2004.
[7] Australian Uranium. Website. (Accessed 16 February 2016, as: http://www.australianuranium. com.au/links.html)

15 Americium

15.1 Introduction

Americium, like all the post-uranic or trans-uranic elements, has been 'discovered' or, more properly, synthesized as part of the research that began with the Manhattan Project. The first recognition that the element had been made was as the result of an experiment in which plutonium was bombarded with neutrons in a nuclear reactor facility in 1944.

More than one isotope of americium exists, and yet it is ^{241}Am that is currently the only isotope that has a widespread use – in household smoke detectors. When compared to the other metals discussed in this book, the total amount of americium in use is small enough that it is not tracked by the United States Geological Survey [1]. However, the World Nuclear Association does keep track of americium use [2].

15.2 Refining and isolation

Production of americium can be shown as a series of nuclear transformations. Starting with ^{238}U, the isotope ^{239}Pu can be synthesized. This is shown in Scheme 15.1.

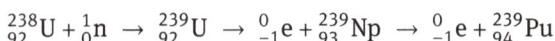

$$^{238}_{92}U + {}^{1}_{0}n \rightarrow {}^{239}_{92}U \rightarrow {}^{0}_{-1}e + {}^{239}_{93}Np \rightarrow {}^{0}_{-1}e + {}^{239}_{94}Pu$$

Scheme 15.1: Plutonium-239 production for 241-americium synthesis.

The series of reactions in Scheme 15.1 is essentially the capture of a neutron, followed by two beta particle emissions (remember that a beta particle is essentially a high energy electron) [3]. Scheme 15.2 shows the subsequent reactions that yield americium-241.

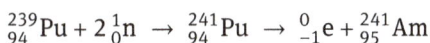

$$^{239}_{94}Pu + 2\,{}^{1}_{0}n \rightarrow {}^{241}_{94}Pu \rightarrow {}^{0}_{-1}e + {}^{241}_{95}Am$$

Scheme 15.2: Americium-241 production.

This represents the capture of two neutrons followed by a beta particle emission. These two schemes represent the pathway that results in usable americium-241, but does not show its separation from other elements. The form in which it is used is AmO_2. This must be separated from other metals and metal oxides, which often comes from what is called spent reactor fuel. The separation of americium from this environment is difficult to represent in terms of reactions, because they are never stoichiometric. Also, some steps must be performed numerous times, especially that for the separation of curium and americium. In broad terms, isolation of americium follows these steps:

1. Metal oxides (sometimes called MOX) are dissolved in nitric acid.
2. Tributyl phosphate is used to remove uranium and plutonium. This is done in a hydrocarbon solvent.
3. Lanthanides are separated using a diamide-based extraction process. This step routinely produces metals (both lanthanides and actinides) in the 3+ oxidation state.
4. Using chromatography and/or centrifugation, americium-containing materials are then extracted from the mixture.
5. Curium and americium are separated from each other as hydroxides, at high temperature, using sodium bicarbonate. This step oxidizes americium to the 4+ state, solubilizing it.
6. Americium oxide (AmO_2) can be reduced using thorium or lanthanum as the reducing agent.

Despite the final step producing americium metal, as mentioned, the use for it is in the AmO_2 form.

15.3 Uses

15.3.1 Smoke detectors

A smoke detector with americium in it has a pathway by which alpha particles move from the source to a detector. When smoke particles are in the air and interfere with these alpha particles, an alarm sounds.

Each household smoke detector contains approximately 0.29 μg (0.00000029 g) of ^{241}Am in the form of americium dioxide (AmO_2). The World Nuclear Association states at its website: "Americium dioxide, AmO_2, was first offered for sale by the US Atomic Energy Commission in 1962 and the price of US$ 1,500 per gram has remained virtually unchanged since. One gram of americium oxide provides enough active material for more than three million household smoke detectors [2]."

15.3.2 Other uses

Americium has found some uses in medical testing and diagnostic equipment. Proposals have also been made to use americium in what are called radioisotope thermoelectric generators on some future spacecraft. Such devices produce both heat and electricity, and may find use in unmanned space probes, but have not yet done so.

15.4 Recycling

The United States Environmental Protection Agency provides basic guidelines for the proper disposal of household smoke detectors, but broad guidelines are not step-by-step instructions [4]. Ultimately, users are asked to contact local agencies and to follow local laws when disposing of these items.

Bibliography

[1] USGS Mineral Commodity Summaries. (Downloaded as: http://minerals.usgs.gov/minerals/pubs/mcs/2015/mcs2015.pdf).

[2] World Nuclear Association. Website. (Accessed 16 April 2015, as: http://www.world-nuclear.org/ and as: http://www.world-nuclear.org/info/Non-Power-Nuclear-Applications/Radioisotopes/Smoke-Detectors-and-Americium/).

[3] Nuclear Energy Institute. Website. (Accessed 13 November 2015, as: http://www.nei.org/).

[4] US EPA. Website. (Accessed 3 June 2015 as: http://www.epa.gov/radiation/sources/smoke_dispose.html).

16 Mercury

16.1 Introduction

Mercury has been known since ancient times, was of interest to western alchemists for centuries, and has been used to refine and isolate silver on a large scale for over five hundred years. It is still recovered from mature silver mines in the United States today [1]. This element has an old, rather romantic name – quicksilver – because it is the only metal that exists as a liquid at ambient temperature.

It is widely known that throughout history, mercury was used to isolate gold because it amalgamates with the yellow metal readily. On the second of Columbus' voyages, a container of mercury was part of one of the ship's cargoes, because it was believed precious metals would be found at the end of the voyage, and mercury might be needed to help extract them [2]. The Spanish metallurgists of the day knew this technique well, since the Almadén mines in Spain had a long history of providing the raw cinnabar ore from which the mercury was extracted.

Today mercury has acquired a rather notorious name because of its toxicity, although the term 'mercury poisoning' is used loosely in the popular press, and thus can indicate two different types of poisoning. Historically, elemental mercury has been used in the production of felt hats, as well as in small-scale gold panning, and the inhalation of vaporous mercury by hatters and by gold miners causes a poisoning that is still sometimes called 'Mad Hatter's Disease', and that is the origin of the expression 'mad as a hatter'. Lewis Carroll's character in the book *Alice In Wonderland*, the Mad Hatter, is based on this.

Contact with organo-mercury compounds in which mercury exists in the +2 oxidation state, especially dimethyl mercury, can cause a much more severe form of mercury poisoning. The death of Professor Karen Wetterhahn of Dartmouth University in 1997, a world expert in heavy metal exposure, is one of the most notable deaths related to this form of mercury exposure. A few drops of dimethyl mercury in solution passed through protective gloves she was wearing, and caused her death within ten months of exposure.

16.2 Refining and isolation

There is more than one type of mineral ore that contains mercury, although cinnabar (HgS) has proven to be the only one economically feasible from which to refine the metal. Livingstonite ($HgSb_4S_8$) and corderoite ($Hg_3S_2Cl_2$) are two other mercury-containing minerals, but are scarce enough that they have not been refined on a large scale. The USGS Mineral Commodity Summary tracks worldwide production of mer-

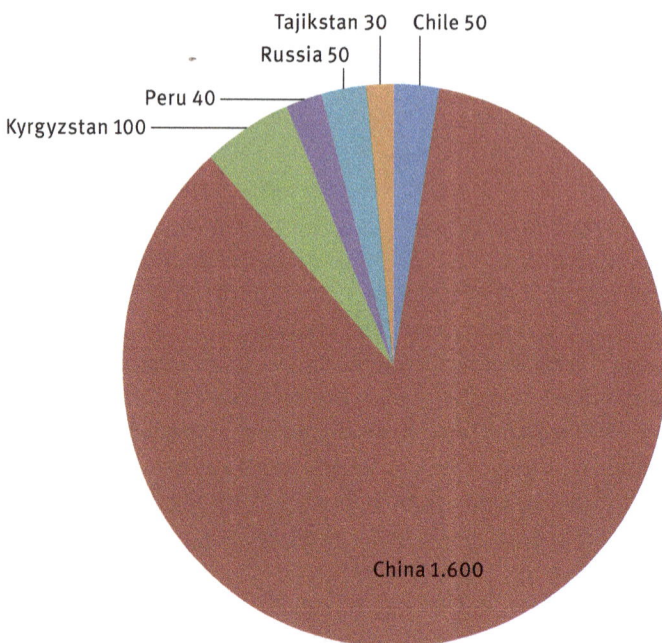

Fig. 16.1: Mercury production, in metric tons.

cury [1], but does not delineate ores types by country. Figure 16.1 shows mercury production globally.

It is plain that China currently dominates the current world production of mercury, but it should be noted that production means new mercury only. Mercury is recycled so completely from large-scale processes that the United States for example,
which has not mined and refined mercury for more than two decades, can still be considered a mercury exporter despite its having enacted the Mercury Export Ban Act of
2008, in Congress [3]. Additionally, Figure 16.1 shows that mercury is produced widely
throughout the world, but does not indicate reserves that have not been worked. Almadén, Spain, for example, has the largest cinnabar reserves in the world. This has
been mined since ancient times, first as a material for pigments, and later for mercury
for amalgamating gold and silver when those two precious metals are refined, but has
not been worked since the year 2000.

The refining of mercury from cinnabar ore can be shown as a simple reaction, in
Scheme 16.1.

$$HgS_{(s)} + O_{2(g)} \rightarrow Hg_{(g)} + SO_{2(g)}$$

Scheme 16.1: Mercury refining.

The reaction chemistry does not show that the ore must first be crushed and concentrated, since not all ores have the same amount of cinnabar in it. As well, since the reaction is run at high temperature, the mercury must be condensed from vapor back to liquid form. Scheme 16.1 could just as correctly show mercury in the liquid state as the product.

16.3 Uses

Despite its toxicity, mercury has found several uses, including one on a large, industrial scale.

16.3.1 Chlor-alkali process

The chlor-alkali process produces sodium hydroxide as its major commercial product, but also produces elemental chlorine gas as well as hydrogen gas. The reaction chemistry is shown in Scheme 16.2.

$$2\,NaCl_{(aq)} + 2\,H_2O_{(l)} \rightarrow 2\,NaOH_{(aq)} + Cl_{2(g)} + 2\,H_{2(g)}$$

Scheme 16.2: Chlor-alkali process.

The scheme does not show the reaction conditions, or the reaction setup. There are currently three different types of process setup: the membrane cell process and the diaphragm cell process, which do not require mercury, and the third type which does require elemental mercury, aptly named the mercury process. In this, the mercury functions as an electrode, the cathode, where it amalgamates the sodium for a brief time before it is re-oxidized and combined to form sodium hydroxide in what is called a decomposer, where hydrogen is also generated and separated [4]. The mercury is in a closed loop that is then recycled back to the cathode.

Mercury emissions from existing chlor-alkali plants remain low – just under ¾ gram measured against each ton of chlorine produced – but European plants using this process are scheduled to be phased out by the year 2020 in favor of plants using the membrane cell process [4]. Beyond those plants using this process in Europe, there are still two using it in the eastern United States [1].

16.3.2 Barometers and thermometers

Mercury thermometers remain perhaps the one end-user item that is most commonly associated with mercury, although the total volume of the metal required for them is small enough that this is not a major use.

16.3.3 Amalgams and compounds

As an amalgam with other metals, mercury has been used for dental fillings for the past century. In the more recent past, various plastic dental filling composites have been marketed, making the market for mercury-containing amalgams smaller.

Batteries containing mercury, known either as mercury–zinc or mercury-oxide cells, have been used in numerous applications where small batteries are required. For example, camera batteries utilized this couple for many years. However, other battery couples that are less toxic have been manufactured on a larger scale in the recent past. Scheme 16.3 shows the reaction chemistry involved in a mercury–zinc battery.

$$HgO_{(s)} + Zn_{(s)} \rightarrow Hg_{(l)} + ZnO_{(s)}$$

Scheme 16.3: Mercury–zinc battery chemistry.

This electrochemical cell provides 1.35 V, and has a working life of over a decade. It functions well in all but extreme cold conditions, but as mentioned, is being replaced by other couples that do not contain mercury.

Mercury salts have also been valued throughout history for their bright colors, and thus have been used in various pigments. The most famous is perhaps the bright red pigment vermillion, which contains crushed cinnabar. This has been almost completely phased out as other, less toxic compounds have been found that impart the same or nearly same colors.

16.4 Possible substitutes

In the chlor-alkali industry, the mercury process is being replaced by the two, competing processes as new plants are brought on line. As mentioned, different amalgams and different battery couples have substituted for mercury-containing ones in the past twenty years.

16.5 Recycling

Almost all mercury is recycled, even from end-user items. The USGS Mineral Commodity Summaries state: "Secondary, or recycled, mercury was recovered from batteries, compact and traditional fluorescent lamps, dental amalgam, medical devices, and thermostats, as well as mercury-contaminated soils [1]." Further, it comments: "Mercury-containing automobile convenience switches, barometers, compact and traditional fluorescent lamps, computers, dental amalgam, medical devices, thermostats, and some mercury-containing toys were collected by as many as 50 smaller companies and shipped to the refining companies for retorting to reclaim the mercury [1]." Clearly, mercury is recycled far more completely than many other metals and materials. Apparently, the only mercury that is not recycled is that in end-user items that individuals choose to discard [5].

In the next five years, decisions will have to be made on what to do with mercury that has been phased out of chlor-alkali plants. A recent article in the trade magazine, *Chemical & Engineering News*, about the use and phase out of mercury in these plants, comments: "...disposing of hundreds of tons of mercury that collectively reside in Europe's chlor-alkali cells will not be easy... The preferred form of internment identified in the EC's best available techniques... is reaction with sulfur and disposal in unused German salt mines [6]." Since this has not yet occurred, one can speculate as to whether there might be some better, safer alternative method.

Bibliography

[1] USGS Mineral Commodity Summaries. (Downloaded as: http://minerals.usgs.gov/minerals/pubs/mcs/2015/mcs2015.pdf).
[2] Thibodeau AM, Killick DJ, Ruiz J, Chesley JT, Deagan K, Cruxent JM, Lyman W. The strange case of the earliest silver extraction by European colonists in the New World, *Proc. Natl. Acad. Sci.*, 2007, 104(9): 3663–3666. Website. (Accessed 14 November 2015, as: http://www.ncbi.nlm.nih.gov/pmc/articles/PMC1805524/).
[3] Mercury Export Ban Act of 2008. (Accessed 5 April 2016, as: https://www.govtrack.us/congress/bills/110/s906/text).
[4] EuroChlor. Website. (Accessed 14 November 2015, as: http://www.eurochlor.org/the-chlorine-universe/how-is-chlorine-produced/the-mercury-cell-process.aspx).
[5] Overarching Framework UNEP Global Mercury Partnership. Website. (Accessed 15 November 2015, as: http://www.unep.org/chemicalsandwaste/Portals/9/Mercury/Documents/Overarching%20Framework.pdf).
[6] A Hurried Good Bye To Mercury: Europe's chemical makers scramble to end mercury-based chlor-alkali production, *Chemical & Engineering News*, November 2, 2015, pp. 24–27.

17 Lanthanides

17.1 Introduction

The lanthanides, sometimes called the lanthanoids, inner transition elements, or the older but still widely used name, the rare earth elements (REEs), were largely discovered and determined to be elements throughout the nineteenth century, with four (promethium, europium, ytterbium, and lutetium) being reported in the early part of the twentieth. One of the difficulties with claiming the discovery of specific lanthanide elements is the same problem that concerns the isolation and large-scale purification of them today – the difficulty of separating two or more from a single ore batch. The lanthanide elements are all very close to each other both in terms of their size and their electronegativity, making them chemically quite similar. This in turn means their separations can be difficult, requiring both time and energy. Because of this challenge, some of the lanthanides were not cleanly separated until the 1950s.

17.2 Refining and isolation

While the lanthanides are chemically similar to each other and to the element yttrium, they do not all ever occur in a single mineral. There are however a large number of minerals that contain one or more lanthanide elements. A non-exhaustive list is shown in Table 17.1.

Several points can be discerned from Table 17.1.

First, several of these minerals appear to contain no rare earth elements. Each formula given is the accepted one for a mineral, but often one or more impurities exist within the structure, often in a relatively large amount, and thus rare earth elements are present as those impurities.

Second, several of the entries have ore names akin to rare earth elements, and several of the elements have names related to the town of Ytterby, Sweden. The minerals and elements are of course related, but the reason for the connection to Sweden is not always clear. In an interesting coincidence, at a time when European scientists were making great strides and advancing knowledge on numerous fronts, it was found that in ores in the mine near Ytterby, Sweden seven of the fourteen rare earth elements were present. Thus yttrium, ytterbium, terbium and erbium all have names derived from the town. Gadolinium, holmium, and thulium are all related to these ores, as well.

Third, more than one of these minerals also contain the element thorium. Thorium's refining and uses were discussed in Chapter 14, and its refining is generally done in conjunction with the refining of the associated rare earths. In many cases, the economic driving force in an ore's purification is the rare earth element or elements, with the thorium being a secondary product. Yet both can become economically profitable to isolate if they can be concentrated sufficiently.

Table 17.1: Minerals containing one or more lanthanide elements.

Ore	General formula	Geographic location	Comments
Aeschynite	$(Nd, Ce, Ca, Th)(Ti, Nb)_2(O, OH)_6$	China, Inner Mongolia, USA	Sources can be high in Ce, Nd, or Y.
Anatase	TiO_2	USA, France	REE within
Ancylite	$Sr(Ce, La)(CO_3)_2(OH) \cdot H_2O$	USA	Source of Ce or La
Apatite	$Ca_5(PO_4)_3(F, Cl, OH)$	Apatity, Russia; Florida, USA; Canada	Can be found without rare earths.
Bastnasite	$(Ce, La, Y)CO_3F$	Sweden, Pakistan, USA	Significant source of Ce
Brannerite	$(Ca, Y, Ce, U)(Ti, Fe)_2O_6$	USA	Source of U
Britholite	$(Ce, Ca)_5(SiO_4, PO_4)_3(OH, F)$	USA	
Brockite	$(Ca, Th, Ce)PO_4 \cdot H_2O$	Colorado, USA	
Cerianite	$(Ce, Th)O_2$	USA	As Ce^{4+}
Cerite	$(Ce, Ca, La)_9(Mg, Fe^{3+})(SiO_4)_6(SiO_3OH)(OH)_3$	Vastmanland, Sweden; Mountain Pass, California, USA; Kola, Russia	Cerite-(Ce) and cerite-(La)
Cheralite	$(Ca, Ce, Th)(P, Si)O_4$	USA	Also recoverable Th
Chevkinite	$(Ca, Ce, Th)_4(Fe, Mg)_2(Ti, Fe)_3Si_4O_{22}$	USA	
Churchite	$YPO_4 \cdot H_2O$	USA	Can also contain REEs
Crandallite	$CaAl_3(PO_4)_2(OH)_5 \cdot H_2O$	USA	
Doverite	$YcaF(CO_3)_2$	USA	
Eudialyte	$Na_4(Ca, Ce)_2(Fe, Mn, Y)ZrSi_8O_{22}(OH, Cl)_2$	USA	Also contains, U, Nb, Ta, Hf
Euxenite	$(Y, Ca, Ce, U, Th)(Nb, Ta, Ti)_2O_6$	USA, Norway	
Fergusonite	$YNbO_4$	USA	Fergusonite-(Y) and -(Ce)
Fluocerite	$(La, Ce)F_3$	Sweden; Kazakhstan; Australia; Inner Mongolia, China	Fluocerite-(La) and fluocerite-(Ce)
Fluorapatite	$(Ca, Ce)_5(PO_4)_3F$	USA	
Fluorite	CaF_2	Very widespread	Y, Yb, and Eu in fluorite often account for the fluorescence
Gagarinite	$NaCaY(Cl, F)_6$	USA	
Gerenite	$(Na, Ca)_2Y_3Si_6O_{18} \cdot 2 H_2O$	USA	REEs as well as Y
Gorceixite	$BaAl_3[(PO_4)_2(OH)_5] \cdot H_2O$	USA	

Table 17.1: (Continued).

Ore	General formula	Geographic location	Comments
Goyazite	$SrAl_3(PO_4)_2(OH)_5 \cdot H_2O$	USA, France	
Hingganite	$(Y, Yb, Er)_2Be_2Si_2O_8(OH)_2$	USA	
Iimoriite	$Y_2(SiO_4)(CO_3)$	USA	
Kainosite	$Ca_2(Y, Ce)_2Si_4O_{12}(CO_3) \cdot H_2O$	USA, Norway	
Loparite	$(Ce, Na, Ca)(Ti, Nb)O_3$	USA Russia	
Monazite	$(La, Ce, Pr, Nd, Y, Th)PO_4$	India; Madagascar; South Africa; Bolivia; Australia	Four different types
Orthite (aka. Allanite)	$(Ce, Ca, Y, La)_2(Al, Fe^{3+})_3(SiO_4)_3(OH)$	Greenland; Queensland, Australia; New Mexico, USA	Designated allanite-(Ce), allanite-(La) or allanite-(Y).
Parasite	$Ca(La, Ce)_2(CO_3)_3F_2$	Colombia; Greenland	Can contain Nd
Perovskite	$CaTiO_3$	USA, Russia, Sweden	May contain Nb
Pyrochlore	$(Na, Ca)_2Nb_2O_6(OH, F)$	Norway	May contain REE and transition metals
Rhabdophane	$(Ce, La)PO_4 \cdot H_2O$	USA	Rhabdophane-(Ce) and -(La), may contain Nd
Rinkite	$(Ca, Ce)_4Na(Na, Ca)_2Ti(Si_2O_7)_2F_2(O, F)_2$	Russia, Greenland	
Samarskite	$(YFe^{3+}Fe^{2+}U, Th, Ca)_2(Nb, Ta)_2O_8$	USA, Russia	
Stillwellite	$(Ca, Ce, La)BSiO_5$	Queensland, Australia; Tajikstan; Ontario, Canada	
Synchysite	$Ca(Ce, La)(CO_3)F$	USA	Synchysite-(Ce) and -(Y) and -(Nd)
Thalenite	$Y_3Si_3O_{10}(F, OH)$	USA	Can occur in zircon
Thorite	$(Th, U)SiO_4$	Norway, USA	
Titanite	$CaTiSiO_5$	Very widespread	Fe, Al, Ce, Y, and Th can be present
Uraninite	$(U, Th, Ce)O_2$	USA, Germany	Usually mined for Th
Vitusite	$Na_3(Ce, La, Nd)(PO_4)_2$	USA	Vitusite-(Ce)
Wakefieldite	$(L, Ce, Nd, Y)VO_4$	Canada; Congo	Four types, based on dominant rare earth
Xenotime	$(Y, Yb, Dy, Er, Tb, U, Th)PO_4$	Brazil; Norway	
Ytterbite (aka. Gadolinite)	$(La, Ce, Nd, Y)_2FeBe_2Si_2O_{10}$	Norway; Sweden; Colorado, USA	Gadolinite-(Y) or gadolinite-(Ce)
Yttrofluorite	$(Ca, Y)F_2$	USA, Sweden	May also contain Tb
Zircon	$ZrSiO_4$	Australia	May contain traces of Hf, U, Th.

Table 17.2: Rare earth element abundances. Non-rare earth elements are in bold.

Element	Symbol	Abundance (ppm)	Atomic #	Year of discovery
Zinc	**Zn**	**75**	**30**	Ancient
Cerium	Ce	68	58	1803
Copper	**Cu**	**51**	**29**	Ancient
Neodymium	Nd	33	60	1885
Lanthanum	La	32	57	1839
Yttrium	**Y**	**30**	**39**	1828
Cobalt	**Co**	**21**	**27**	1735
Lead	**Pb**	**20**	**82**	Ancient
Samarium	Sm	20	62	1879
Gadolinium	Gd	20	64	1886
Praseodymium	Pr	16	59	1885
Dysprosium	Dy	13	66	1950s
Ytterbium	Yb	10	70	1953
Hafnium	Hf	8	72	1923
Erbium	Er	7	68	1934
Tin	**Sn**	**3**	**50**	Ancient
Holmium	Ho	3	67	1878
Terbium	Tb	3	65	1843
Europium	Eu	2	63	1890
Lutetium	Lu	2	71	1907
Thulium	Tm	0.7	69	1911
Uranium	**U**	**0.03**	**92**	1789

Fourth, many of these minerals have multiple rare earth elements in them. This is because of the similarity of chemical reactivity among these elements. While this can be advantageous, it can also be a problem when the separation of the rare earths is required.

Interestingly, although the old name 'rare earth element' is still widely applied to the lanthanides, some are not particularly rare. Table 17.2 shows a listing of the REEs along with several other elements. Note that tin is less common than over half of the REEs, and uranium is less common than virtually all of them. The moniker 'rare earth element' is simply an historical artifact, since none of these elements occurs in any large concentration.

Production of the rare earth elements is tracked by several organizations, including the USGS Mineral Commodity Summaries each year, as well as by the US Department of Defense Strategic and Critical Materials 2013 Report on Stockpile Requirements, the US Department of Energy Critical Materials Strategy, and the British Geological Survey, as well as several other national and international organizations [1–9]. A global breakdown of their production is shown in Figure 17.1 in what is called rare earth oxide (REO) equivalents, in metric tons. After a long hiatus, one company in the United States is again mining rare earths. Rare earths had been mined in the United

Vietnam 200

USA 7.000   Australia 2.500

Thailand 1.100

Russia 2.500

Malaysia 200

India 3.000

China 95.000

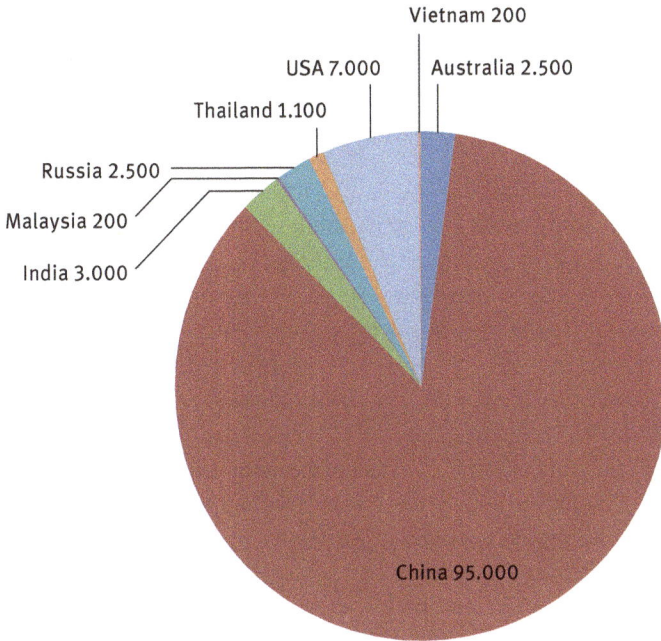

Fig. 17.1: Rare earth oxide production, in metric tons.

States up until 1992, but the operation was discontinued because prices on world markets had at the time dropped low enough that it was not profitable. This and the growing uses for rare earth elements shifted the supply source from the United States as a major producer to China being by far the largest producer.

It is clear that China currently dominates the world's supply of the rare earth elements. What is not obvious from Figure 17.1 is that the figure for China may actually be somewhat low. Several news reports in the past five years have discussed the illegal export and trade in rare earth oxides from China, especially as the price for them has risen on world markets.

Refining of the rare earth elements is difficult to draw in terms of reaction chemistry, because most steps do not involve stoichiometric amounts of reactants or products, and because rare earth elements do not all occur in any one ore batch. But broad steps for their separation can be listed as follows:

1. Milling and grinding. As with many different elements, ores must first be milled and ground, usually to the consistency of powder. The idea is to homogenize the sample and maximize surface area of the raw ore, so that subsequent chemical and physical processing steps take place as completely as possible.
2. Electromagnetic separation. Some REE ores are magnetic, and thus a separation step involves placing ores on conveyor belts that have magnets in the end rollers. When the ore powder passes over the end of the roller, while it all ultimately falls

off, non-magnetic ore falls directly off the end, while magnetic ore remains attached to the belt until it has advanced far enough that the belt and roller (with magnet in it) again separate, at which point the magnetic ore falls.

3. Ore flotation. The process known as ore flotation or froth flotation was designed to help concentrate copper ores, since like rare earth ores, they can exist with very small amounts of the desired ore in a large amount of silicates and other unneeded material. The step involves the separation of particles based on their hydrophobicity. It usually involves a surfactant additive to help separate different components, and has been used extensively with sulfide ores.

4. Centrifugal concentration. Using centrifuges to help concentrate ores is a matter of separating the components based on small differences in densities. Depending on the ore batch, this process may need to be repeated numerous times, with each cycle slightly furthering the separation.

5. Leaching and precipitation. This step can incorporate three different techniques:
   (a) Fractional crystallization. Selective precipitation of materials is effected by the control of both pH and temperature. This results in materials with lower solubility crystallizing from the mixture.
   (b) Ion exchange. This is a hydrometallurgical process that involves separating components of a mixture using a resin to attract one fraction of the mixture, while the other passes through. Rare earths that are bound to the resin are then washed from it with various solvents or solvent mixtures.
   (c) Solvent-based extraction. Two or more rare earths can be separated using a two-phase solvent system in which the rare earth elements have differing affinities for the two solvents. The solvents are not miscible with each other, thus allowing the rare earths to be concentrated each in one of the two liquids. This can be combined with changing the pH of the solutions to effect precipitation of one element preferentially.

6. Electrolytic deposition. Like most metals, rare earths can be electrodeposited from solutions. This can start with a solution containing the rare earth element(s) and an anode that is inert, or can begin with recycled material functioning as the anode [14].

Each of these broad steps may need to be repeated for complete separation. In the past, when methods had not yet been improved, some of these separation steps were performed hundreds or even thousands of times.

17.3 Uses

Rare earth elements find several industrial uses, and the demand for them has been growing swiftly in the last ten years. As mentioned, their production is tracked by different organizations, and the USGS Mineral Commodity Summaries states the following concerning their use: "The estimated distribution of rare earths by end use was as follows, in decreasing order: catalysts, 60 %; metallurgical applications and alloys, 10 %; permanent magnets, 10 %; glass polishing, 10 %; and other, 10 %." Clearly, catalysts are the largest single category of use. We will discuss them briefly, but focus our discussion on alloys and magnets, since these are both uses for the reduced elements.

17.3.1 Catalysts

Cerium finds use as an automotive catalyst in greater amounts than the other rare earths, usually as CeO_2. Called ceria, this compound can be made by calcining other cerium compounds at high temperatures. In turn, these other compounds can be produced from the elemental metal. Scheme 17.1 shows a representative synthetic pathway, although the reactions are not balanced.

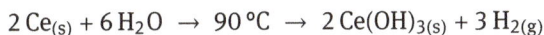

$$2\,Ce_{(s)} + 6\,H_2O \rightarrow 90\,°C \rightarrow 2\,Ce(OH)_{3(s)} + 3\,H_{2(g)}$$

followed by

$$Ce(OH)_{3(s)} \rightarrow CeO_{2(s)} + H_2O_{(g)}$$

Scheme 17.1: Production of ceria.

Ceria is used in conjunction with precious metals catalysts such as platinum to increase the efficiency of the catalyst. In a catalytic converter, ceria is referred to as the 'washcoat', which enhances the surface area, and thus the amount of the precious metal that actively catalyzes gas transformations as they pass through the converter.

17.3.2 Magnets, $Nd_2Fe_{14}B$, plus $SmCo_5$ and Sm_2Co_{17}

Neodymium has found extensive use in the past decade in what are sometimes called super magnets. A variety of different websites detail how such magnets are made, but also differ in some details. This may be because of proprietary concerns on the part of manufacturers. In general though, the steps in manufacturing neodymium magnets are as follows:
1. Mixing the components as metal powders and melting them in a furnace.
2. The resulting alloyed ingot is ground to a fine powder.
3. In the presence of a magnetic field supplied by a solenoid, the alloy powder is pressed.
4. The object is sintered (heat treated below its melt point) under argon to give strength to the newly-formed magnet and to prevent a de-magnetized oxide 'skin' from forming. This tends to shrink the resulting magnet slightly.

5. Machining. Magnets must be machined to the correct size for each specific application.
6. Plating. Often, magnets are plated with another metal or alloy to enhance their corrosion resistance [10].

The actual composition of what get called 'neodymium magnets' is not pure neodymium. Rather, it is an alloy of neodymium, iron, and boron, often with small amounts of aluminum, dysprosium, and niobium added. Alloy compositions can be proprietary, but general percentages are as follows: Fe 64.5–68.5%, Nd 29–32%, B 1–1.5%. The other three elements in the alloy are routinely below 1% each.

Samarium–cobalt magnets are of two general types, often referred to as Series 1:5 or Series 2:17 based on their composition. Their production generally follows the steps listed above, with alloying, crushing, pressing, sintering, and machining. This results in anistropic magnets (meaning magnets with an axis of orientation). These magnets are more brittle than neodymium magnets, and care must be taken when they are brought in contact with each other.

17.3.3 Alloys

As with many elements, small amounts of rare earth elements can be used in alloys to produce metals with desired properties. There is no underlying theory for alloy production. Rather, trial and error essentially is used, and ends when an alloy with suitable properties for a specific application are found. Table 17.3 shows several alloys that incorporate rare earth elements. Neodymium is listed, although its use in alloys is essentially that which was already discussed, magnets.

Table 17.3: Alloys containing rare earth elements.

REE(s)	Alloyed with	Major use(s)	Comments
Scandium	Aluminum	Baseball bats, sporting equipment	
Terbium or dysprosium	Iron	Speakers	Sometimes called 'terfenol-D'
Cerium 50 % and lanthanum 25 %, neodymium, praseodymium	Iron oxide, magnesium oxide	Ignition device in lighters	Sometimes called 'ferrocerium'
Cerium	Aluminum	Lightweight equipment	
Neodymium	Iron and boron	Magnets	

17.3.4 Heavy rare earth element uses

Several of the lighter rare earth elements, as well as the elements scandium and yttrium, have been discussed in detail, because there are numerous uses for them in an ever-widening series of applications. The final five elements in the lanthanide period – holmium, erbium, thulium, ytterbium, and lutetium – currently have no major, large-scale use.

17.4 Possible substitutes

Magnets containing rare earth elements are very powerful and effective, and substitutes for them are being sought. For several years in the recent past, the United States National Science Foundation has put out calls specifically for research proposals aimed at developing materials of equal or greater magnetic capability to the REEs, and yet which do not incorporate them. There have not yet been press releases or other announcements indicating success in this field that has been translated to commercial production of magnets [11–14].

17.5 Recycling

The recycling of rare earth elements from various end-user items is in its infancy today. In some malls and phone stores throughout the United States, there are kiosks at which old cell phones can be turned in for recycling. This perhaps obviously includes recycling the neodymium-based magnet as well as other components. But no state or municipality yet appears to have passed any law about the recycling of items that contain rare earth elements.

Bibliography

[1]   USGS Mineral Commodity Summaries. Downloaded as: http://minerals.usgs.gov/minerals/pubs/mcs/2015/mcs2015.pdf
[2]   US Department of Defense Strategic and Critical Materials 2013 Report on Stockpile Requirements. Downloadable as: http://mineralsmakelife.org/assets/images/content/resources/Strategic_and_Critical_Materials_2013_Report_on_Stockpile_Requirements.pdf
[3]   US Department of Energy Critical Materials Strategy. Downloadable as: http://energy.gov/sites/prod/files/DOE_CMS2011_FINAL_Full.pdf
[4]   British Geological Survey, Rare Earth Elements. (Accessed 5 April 2016, as: www.MineralsUK.com).
[5]   Canadian Rare Earth Element Network. Website. (Accessed 19 November 2015, as: http://www.cim.org/en/RareEarth/Home/AboutUs.aspx).

[6] The major rare-earth-element deposits of Australia: geological setting, exploration, and re-sources. Downloadable at: http://www.ga.gov.au/corporate_data/71820/Complete_Report.pdf

[7] United Nations, GSDR 2015 Brief, Rare Earth Elements; From Mineral to Magnet. Download-able at: https://sustainabledevelopment.un.org/content/documents/5749Rare%20earth%20elements,%20from%20mineral%20to%20magnet.pdf

[8] European Commission. European Rare Earths Competency Network (ERECON). Strengthen-ing the European Rare Earths Supply Chain. Downloadable at: http://ec.europa.eu/growth/sectors/raw-materials/specific-interest/erecon/index_en.htm

[9] RARE, The Association for rare Earths. Website. (Accessed 16 April 2015, as: http://www.rareearthassociation.org/).

[10] US Magnetic Materials Association. Website. (Accessed 16 April 2015, as: http://www.usmagneticmaterials.com/).

[11] Congressional Research Service. Rare Earth Elements: The Global Supply Chain. Website. (Ac-cessed 16 April 2015, as: https://www.fas.org/sgp/crs/natsec/R41347.pdf).

[12] Congressional Research Service. Rare Earth Elements in National Defense: Background, Oversight Issues, and Options for Congress. Website. (Accessed 16 April 2015, as: https://www.fas.org/sgp/crs/natsec/R41744.pdf).

[13] NdFeB-Info.Com. Website. (Accessed 19 November 2015, as: http://www.ndfeb-info.com/neodymium_magnets_made.aspx).

[14] United States Environmental Protection Agency. Rare Earth Elements: A Review of Production, Processing, Recycling, and Associated Environmental Issues. Website. (Accessed 24 November 2015, as: http://nepis.epa.gov/Adobe/PDF/P100EUBC.pdf).

18 Lead

18.1 Introduction

Lead is a post-transition metal that continues to occupy an important place in several industries and end-user applications within modern life, but that has an ancient history as well. A variety of lead-bearing ores exist, and several of them have proven to be quite easy to extract the metal from. In the last few decades, lead recycling has also become a mature industry throughout much of the world, and even relatively small items that contain some lead are in some way dismantled for it and for other recyclable metals.

18.2 History

The atomic symbol for lead – Pb – hints at its ancient roots. The symbol comes from the name 'plumbum', which is the Latin word for the element. Interestingly, this term also is the root word for the modern English words 'plumber' and 'plumbing', because lead was used in the Roman Empire in projects that directed water throughout Rome and some of the more developed cities of that time.

Lead also has something of a sinister reputation, because of what is broadly called lead poisoning. In modern society, this tends to occur when people ingest lead (IV) oxide, the now classic example being children ingesting paint chips. This is because very young children will eat anything that tastes good, and the lead oxide added to paints to give them a specific look while dry that also inadvertently makes them taste sweet. Interestingly, in the ancient world, and elsewhere in history, lead poisoning occurred because of a technique that was developed to sweeten tart wines. A single vat of a tart wine was boiled down to about $\frac{1}{2}$ of its original volume – in a lead container! A single ladle of the resultant liquid material was then added to containers of the same tart wine batch, which did sweeten all the containers of wine. It also caused acute lead poisoning. In what may be the first trade embargo in history, the people referred to as the Germanens and Teutons, east and north of the Rhine River, at times refused to trade for Roman wines and other goods because they knew there was something in them that made them sick. While they certainly did not have the knowledge to determine the cause of their lead poisoning, they were aware of the source of the problem, and took what measures they could to combat it.

18.3 Refining and isolation

Lead is derived from several different minerals, which are shown in Table 18.1. While the percentage of lead in each mineral is listed, it should be noted that such percentages are for a pure mineral, and minerals are themselves not always found pure and unadulterated. Many deposits must be enriched, and thus removal of rock, co-mingled material, and what is sometimes called overburden must take place as the first step in processing the ore.

Figure 18.1 shows the global breakdown of lead production by country, in thousands of metric tons, using data from the USGS Mineral Commodity Summaries. Lead is produced on a large enough scale that there is more than one organization that tracks it [1–5].

Table 18.1: Lead-bearing ores.

Formula	Ore common name(s)	% lead in ore	Geographic location	Comments
PbO	Lead monoxide, litharge, plumbous oxide, massicot	92.9	USA, England	An uncommon ore. Sometimes synthesized when needed.
Pb_3O_4	Lead tetroxide, minium, red lead, triplumbic tetroxide	90.7	Spain	
PbO_2	Lead dioxide, plattnerite, plumbic oxide, scrutinyite	86.7	Europe, Mexico, USA, Russia, Australia, Namibia, Iran	
PbS	Galena, lead (II) sulfide, blue lead, lead glance	86.7	USA, Canada, Germany, Italy, England, Bulgaria, Australia, Israel	Major source of lead for refining, US town named: Galena, Illinois
$PbSO_4$	Angelsite, linarite	68.3	Spain	An uncommon ore
$PbCO_3$	Cerussite, white lead, lead carbonate	77.5	Australia, Germany, USA	Soluble in acids
$PbMnO_4$	Wulfenite, yellow lead	63.5	Austria, USA, Mexico, Slovenia	Forms crystals of variable color
$Pb_5(AsO_4)_3Cl$	Mimetite, green lead	69.7	Mexico, Namibia	Seldom mined exclusively for the lead
$Pb_5(PO_4)_3Cl$	Pyromorphite, green lead	88.9	Australia, Mexico	
$Pb_5(VO_4)_3Cl$	Vanadinite, green lead	73.2	USA, Morocco, Argentina, Namibia	Widely occurring ore

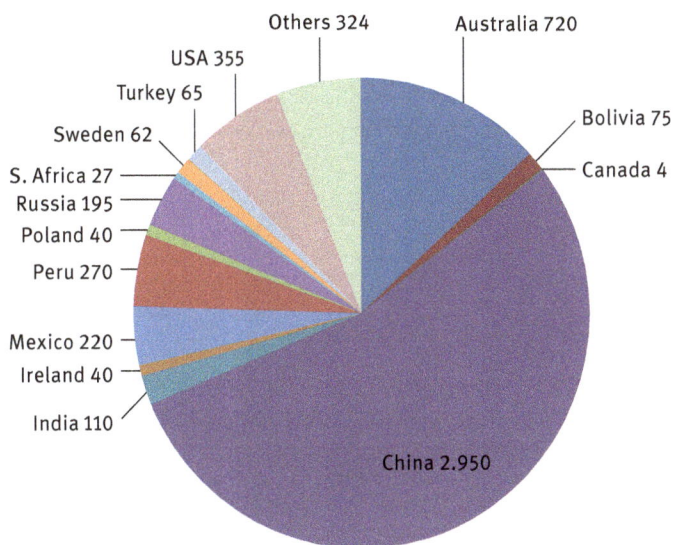

Fig. 18.1: Lead production worldwide, in thousands of metric tons.

Lead is considered a common or base metal, and while it certainly has value, it is considered secondary to some other metals such as silver. Lead currently is co-produced with other metals in many instances, and thus ends up being considered the co-product.

For lead sulfide ores, such as galena, the lead must first be converted to an oxide, then smelted. Scheme 18.1 illustrates the basic chemistry of this process.

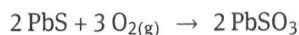

$$2\,PbS + 3\,O_{2(g)} \rightarrow 2\,PbSO_3$$
followed by
$$PbSO_3 \rightarrow PbO + SO_{2(g)}$$
followed by
$$PbO + CO \rightarrow CO_{2(g)} + Pb_{(l)}$$

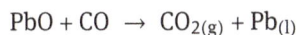

Scheme 18.1: Lead sulfide refining.

As with all metals reduced from oxides, the process is undertaken at high temperature, in this case in what is called a lead smelter. The molten lead is separated from any dross and other material, and can then be formed into ingots. The production of purer lead is usually accomplished using one of three processes: the Parkes, the Betterton–Kroll, or the Betts processes.

The Parkes process

Lead and silver can form solutions at high, molten temperatures. The Parkes process adds zinc to such a solution, because silver is soluble in zinc as well, while lead is not. Scheme 18.2 shows the separation chemistry as a basic reaction, although it is not usually stoichiometric.

$$Pb : Ag_{(l)} + Zn_{(s)} \rightarrow Pb_{(l)} + Zn : Ag_{(l)}$$

Scheme 18.2: The Parkes process.

The molten zinc–silver solution can be heated until the zinc boils, leaving the silver behind. The zinc is then re-condensed by retorting, and re-used. Depending on concentrations, the silver can be the most valuable product, although this is designed to purify lead.

The Betterton–Kroll process

This process is generally used to separate lead from bismuth, with which it is soluble. Scheme 18.3 shows the simplified reaction chemistry. Again, this is not normally a stoichiometric process.

$$Ca + Mg + Pb : Bi_{(l)} \rightarrow Ca : Mg : Bi_{(s)} + Pb_{(l)}$$

Scheme 18.3: The Betterton–Kroll process.

The products are separated based on their density, with molten lead being the denser (the reaction is generally run at around 500 °C), and thus sinking to the bottom. Beyond collecting the lead, the bismuth-containing dross can be purified to extract the bismuth using elemental chlorine.

The Betts electrolytic process

This process has been used since patents were filed for it in 1901, and is primarily used to separate lead from a variety of other metals that have been reduced with it and which remain in the lead when it is cooled. It is an electrolytic process that uses the impure lead as an anode, a starter cathode of extremely pure lead, and an electrolyte that is a mixture of hexafluorosilicic acid (H_2SiF_6) and lead fluorosilicate ($PbSiF_6$). A voltage is applied and the lead anode dissolves into the solution, leaving behind other metals in what is called an anode mud. If the anode mud contains metals such as silver or gold, these metals are also recovered, refined, and used.

By careful control of the voltage in the Betts process, highly pure lead is formed at the cathode. This process tends to be more expensive than the just-mentioned Parkes and Betterton–Kroll processes, and is used only when high purity of the lead is imperative.

18.4 Uses

18.4.1 Batteries

Lead-acid batteries for automobiles and trucks may be the end-user item with which people are most familiar. There are certainly a wide variety of batteries used in the modern world, but the well-established lead-acid battery – first produced in the mid-1800s – functions on chemistry that can be described as rare, since it is lead that undergoes both the oxidation and reduction, while almost all other batteries require the redox of two different metals. Scheme 18.4 shows the basic reaction chemistry.

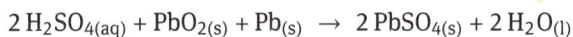

$$2\,H_2SO_{4(aq)} + PbO_{2(s)} + Pb_{(s)} \;\rightarrow\; 2\,PbSO_{4(s)} + 2\,H_2O_{(l)}$$

Scheme 18.4: Lead-acid battery chemistry.

This redox reaction sets up an approximately two volt couple, and thus 12 V car batteries are manufactured by linking six cells in series.

18.4.2 Construction material

Lead has found use for centuries as a roofing material. There exists a Lead Sheet Association that is concerned with how lead sheet is manufactured and used. At their website they note that lead can be used as construction cladding in addition to roofing material, and take pains to point out that some lead roofing and gutter material on European buildings remain serviceable and in use after 500 years [3].

Lead sheet can be produced by rolling, by what is called machine casting, or by sand casting. Sand casting has the longest history, having apparently been practiced in ancient times. In this technique, a bed of sand is smoothed in a form with raised sides, and molten lead is poured on one end. It then runs to the other, where some falls over the far edge and is caught for later use. As it cools, the lead sheet can then be cut and rolled.

18.4.3 Ammunition, hunting and military

Although lead has been a traditional material for making bullets when black powder muskets and rifles were used, in many places lead has been outlawed in hunting ammunition. In such cases, it is replaced by iron or other metals, because they are environmentally more benign (keep in mind that much hunting shot ends up in the environment, since many shots do not hit their target). Similarly, what are called 'green bullets' are being used by the military of several nations, because there is no soil contamination at shooting ranges when lead core ammunition is replaced by some other metal, such as iron.

For hunters who continue to practice what is called black powder hunting, lead balls are still used as a form of ammunition, although this is a much smaller use than for batteries and construction material.

18.4.4 Alloys

Pewter is the subject of Chapter 5, and has traditionally used a small amount of lead. This has been phased out of virtually all modern pewter objects.

Lead combines with tin to form solders with relatively low melting points. Binary mixtures of lead and tin reach a eutectic point at approximately 55 % tin and 176 °C. The presence of any impurities in the lead or tin when such an alloy is formed can depress the melting point as much as a further 80 °C. Thus, impure lead–tin alloys can actually melt in boiling water.

Lead is also added to copper to produce a material with better machinability than copper alone. Such alloys are usually used in bearings in a variety of applications [7]. In cooling, the copper solidifies first, and lead acts as a lubricant, since its melting point is so much lower than that of copper.

18.5 Possible substitutes

In many applications, iron or tin can substitute for lead. Often, the substitution is a matter of the perception that lead may not be environmentally benign. Additionally, plastics such as polypropylene can be used to substitute for lead and other metals in containers and cable sheathing.

18.6 Recycling

Lead recycling is a developed, mature industry, with lead from almost all sources being recycled and handled by scrap yards. Lead-acid batteries are typically recycled for their lead and for the polypropylene casing. The sulfuric acid is so inexpensive that it is routinely neutralized with base and then discarded. Concerning recycling, the International Lead Association states at their website: "Lead enjoys one of the highest recycling rates of all materials in common use today." While this is obviously a trade organization putting the best possible spin on their material, it is true that lead is easy to recycle, in large part because of its low melting point, for a metal, and thus ease of separation from other materials. Lead sheeting from the repair and demolition of buildings is always sent to scrap yards for re-melting and re-use.

Bibliography

[1] USGS Mineral Commodity Summaries. Downloaded as: http://minerals.usgs.gov/minerals/pubs/mcs/2015/mcs2015.pdf

[2] International Lead Association. Website. (Accessed 16 April 2015, as: http://www.ila-lead.org/).

[3] European Lead Sheet Industry Association. Website. (Accessed 16 April 2015, as: www.elsia-web.org).

[4] Lead Development Association International. Website. (Accessed 16 April 2015, as: www.ldaint.org).

[5] Lead Sheet Association. Website. (Accessed 16 April 2015, as: www.leadsheetassociation.org.uk).

[6] Index Mundi. Website. (Accessed 30 November 2015, as: http://www.indexmundi.com/minerals/?product=lead).

[7] Copper Development Association Inc. Website. (Accessed 30 November 2015, as: http://www.copper.org/resources/properties/microstructure/cu_leaded.html).

19 Tungsten

19.1 Introduction

Tungsten is another metal that has a relatively short history, having only been isolated and discovered in 1781, although there had been work with tungsten-containing minerals before this. The atomic symbol 'W' is related to the early work with tungsten ores that occurred when trying to separate it from tin. The name 'wolfram' translates to 'wolf cream' and is rooted in mining history. Historically, whenever the white foam of a tungsten ore formed during tin refining at operations in Germany and Austria, the end result was less tin that could be recovered. To the metallurgists of the day, this was analogous to a wolf devouring some of the sheep – or more precisely, some of the tin metal they were trying to isolate.

Tungsten has become very important in a variety of applications in the modern world, always because of its hardness as an alloy or component of a ceramic. Patents related to hard alloys and ceramics of tungsten go back as far as the middle of the nineteenth century, and the tungsten filament proved to be an excellent material for the new electric lights of the early twentieth century. Also, the knowledge that tungsten alloys were useful to the world's militaries was realized at the beginning of the twentieth century, as armies began to mechanize and use heavily armored vehicles and equipment. The Portuguese sources of tungsten, although not the largest in the world, were very important during World War II, because Portugal remained neutral and was able to deal with and sell to any nation in either the Axis or Allies.

Because there are currently numerous producers of tungsten worldwide, there is now an International Tungsten Industry Association. Its membership includes over 50 companies, many of which are shown in Table 19.1 [1].

It is obvious from Table 19.1 that there are several major tungsten producers in China and the nations of the Pacific Rim, yet there are tungsten producers throughout all six inhabited continents. Australia appears to have several major producers of tungsten and tungsten-related products. Note that the third column remains empty for several of the companies listed, because tungsten appears to be their main and major product.

Additionally, among the other metals associated with tungsten producers, molybdenum appears several times in Table 19.1. The co-location of molybdenum with tungsten on the periodic table is an indicator that molybdenum can sometimes be mined as a component of tungsten-bearing ores, although there are mines in which molybdenum is the primary product. While its use tends to be restricted to steel alloys, there are enough producers that a trade organization also exists that focuses on molybdenum production [2].

Table 19.1 also makes it clear that some companies concentrate their efforts in what is best called hard materials, while other companies either have expanded to

Table 19.1: Tungsten-producing companies.

Name	Headquarters	Other materials produced	Website
A & M Minerals & Metals Ltd	Britain	Bi, Cd, Co, Ge, Pb, Nb, Re, Sb, Ta, Sn	www.amgroup.uk.com
Advanced Material Japan Corp	Japan		http://www.amjc.co.jp/
Almonty Industries Inc	Canada		www.almonty.com
ALMT Corp	Japan	Mo	www.allied-material.co.jp
Amalgamated Metal Corporation Plc	Britain		www.amcgroup.com
Asia Tungsten Products Vietnam Ltd	Vietnam	FeW	www.asiatungsten.com.vn
Atlas Copco Secoroc AB	Sweden		www.atlascopco.com
Betek GmbH & Co KG	Germany		www.simongruppe.de
Blackheath Resources Inc	Canada		www.blackheathresources.com
Brazil Tungsten Holdings Ltd	Brazil		www.braziltungsten.com
Carbine Tungsten Ltd	Australia		www.carbinetungsten.com.au
CB-CERATIZIT CN	China		www.cbceratizit.com
CERATIZIT SA	Luxembourg		www.ceratizit.com
China Minmetals Corp	China	Fe	www.minmetals.com
Chongyi Zhangyuan Tungsten Co Ltd	China	WC	www.zy-tungsten.com
Eurotungstene – Eramet Group	France	Co, Re	www.eurotungstene.com
Federal Carbide Co	USA		www.federalcarbide.com
Global Tungsten & Powders Corp	USA	Mo	www.globaltungsten.com
Grondmet GmbH & Co Kg	Germany		www.grondmet.com
Guangdong Xianglu Tungsten Co Ltd	China	WC, WO_3	www.xl-tungsten.com
Hazelwood Resources Ltd	Australia	FeW	www.hazelwood.com.au
HC Starck GmbH	Germany	Mo, Ta, Ni, Re	www.hcstarck.com
ICD Alloys and Metals LLC	USA	Sb, Cd	www.icdgroup.com
Japan New Metals Co Ltd	Japan	Mo, borides, nitrides	www.jnm.co.jp/eng/index.html
Jiangxi Rare Metals Tungsten Holdings Group Co Ltd	China		www.jxtc.com.cn
Jiangxi Tungsten Industry Group Co Ltd	China	Rare earths, Ta, Nb	www.jwyx.com.cn/eng/default.htm
Jiangxi Yaosheng Tungsten Co Ltd	China	Sn	www.w.jx.cn
Kennametal Inc	USA		www.kennametal.com
King Island Scheelite Ltd	Australia		www.kingislandscheelite.com.au/
Mi-Tech Metals Inc	USA	Ni, Fe, Cu	www.mi-techmetals.com

Table 19.1: (Continued).

Name	Headquarters	Other materials produced	Website
Newcrest Mining Ltd	Australia	Au, Cu	www.newcrest.com.au
Nippon Tungsten Co Ltd	Japan		www.nittan.co.jp/en/index.html
Noble Group – Hard Commodities	Singapore		www.thisisnoble.com
North American Tungsten Corp Ltd	Canada		www.northamericantungsten.com
Ormonde Mining Plc	Ireland		www.ormondemining.com
Premier African Minerals Ltd	British V.I.	Li, Ta, Ni, rare earths	www.premierafricanminerals.com
Sandvik Machining Solutions AB	Sweden	Diamond, carbides	www.sandvik.com
Saxony Minerals & Exploration AG	Germany	Sn, In	www.smeag.de
Sojitz Beralt Tin & Wolfram (Portugal) SA	Portugal	Sn	
Sumitomo Electric Industries Ltd	Japan	Diamond, carbides	www.sumitool.com
TaeguTec Ltd	S. Korea		www.taegutec.co.kr
Thor Mining Plc	Australia	Mo, Au	www.thormining.com
Tikomet Oy	Finland	WC	www.tikomet.fi
Todd Corp	New Zealand	Mo	www.toddcorporation.com
Toshiba Materials Co Ltd	Japan	Ni, Cu, nitrides	www.toshiba-tmat.co.jp/tmat/eng/index.htm
Tranzact Inc.	USA	Mo, Ta, Nb, Cr, Hf, Re	www.tranzactinc.com
Tungco, Inc.	USA	WC, alloys	www.tungco.com
Umicore, SA	Belgium	Zn, Ni, Pt	www.umicore.com
Viet Nam Youngsun Tungsten Industry Co., Ltd.	Viet Nam		www.wfyoungsun.com
Vital Metals, Ltd.	Australia	Au, Zn	www.vitalmetals.com.au
W Resources, Plc.	Britain		www.wresources.co.uk
Wogen Resources, Ltd.	Britain	Bi, ferro-alloys, REEs	www.wogen.com
Wolf Minerals, Ltd.	Australia	Sn	www.wolfminerals.com.au
Wolfram Bergbau und Hutten, AG	Austria	WC	www.wolfram.at
Wolfram Company CJSC	Russia	Mo, Re	www.wolframcompany.ru/wmc/en
Xiamen Tungsten Co., Ltd.	China		www.cxtc.com
Zhuzhou Cemented Carbide Group Corp., Ltd.	China	Mo, Ta, Nb	www.chinacarbide.com
Zigong Cemented Carbide Co., Ltd.	China	Mo	www.zgcc.com

include tungsten within their profile of metal concerns, or have initially started with tungsten, then expanded to include interests in other metals as well.

Worldwide, tungsten production is not only tracked by the ITIA, but by the USGS Mineral Commodity Summaries annually, as well as by the British Geological Survey. Figure 19.1 shows the breakdown by country in metric tons [3, 4].

There are less than ten companies in the United States that produce tungsten, but that are not included in Figure 19.1 for proprietary reasons. It is obvious that China currently dominates tungsten production, but to give a sense of overall scale of this production, consider the following: Tungsten has a density of 19.25 g/cm^3, or $19,250$ kg/m^3, which is much greater than lead, and close to that of gold. This means that a cubic meter of tungsten metal has a mass of 21.22 tons (USA tons or short tons). This in turn means that a national output like that of Portugal would be less than 30 of such one-meter cubes, while the output of China would be just over 3,200 of such cubes. The aim of this comparison is to point out how small the overall output of tungsten is globally, certainly when compared to more common, higher volume metals.

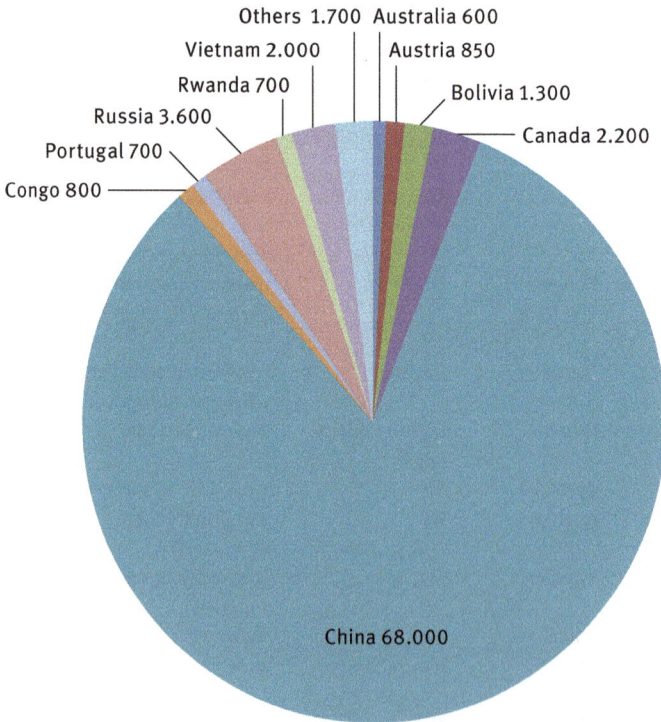

Fig. 19.1: Worldwide tungsten production, in metric tons.

19.2 Refining and isolation

Tungsten does not exist in as many ores as some other metals, although tungsten ores are found on all six inhabited continents. Curiously, tungsten appears to be the only metal found in the anionic portion of all of its ores. The ores from which tungsten can be refined profitably are listed in Table 19.2.

Table 19.2: Tungsten ores.

Ore name	Formula	% W	Comments
Ferberite	$FeWO_4$	60.5	Discovered 1863
Hubnerite	$MnWO_4$	60.7	Discovered 1865
Scheelite	$CaWO_4$	48.3	Named after the discoverer of tungsten. Can be synthesized
Wolframite	$(Fe, Mn)WO_4$	51.3	Variable amounts of iron and manganese

As with many metals, tungsten ores must first be concentrated before refining. And so, concentrating the ore by crushing it to uniform particle size, and separating rock and other non-tungsten containing components, is critical. It is difficult to write reaction chemistry to show the process, but the broad steps are as follows:

1. Gravity and flotation – tungsten ores are dense, and can initially be separated from other materials using these methods.
2. Magnetic separation – wolframite is often magnetic because of the iron component of it, and this can be separated from the scheelite fraction by magnetic means.
3. Tungsten concentrate – the resultant tungsten concentrate can be treated using a concentrated caustic solution (NaOH), resulting in sodium tungstate (Na_2WO_4), which is further purified using filtration and/or precipitation. The reaction for this can be represented as:

$$FeWO_4 + 2\,NaOH + 2\,H_2O \rightarrow Fe(OH)_2 + Na_2WO_4 \cdot 2\,H_2O$$

although it is not normally run stoichiometrically.
4. The sodium tungstate can then be treated with an ammonia solution to produce ammonium paratungstate (APT), with the formula $(NH_4)_{10}(H_2W_{12}O_{42}) \cdot 4\,H_2O$. The APT is then heated to force decomposition, leaving WO_3 [5].

Despite the complexity of the earlier steps, the final step in purifying tungsten metal can be written in a straightforward manner, as shown in Scheme 19.1.

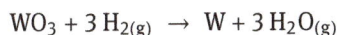

$$WO_3 + 3\,H_{2(g)} \rightarrow W + 3\,H_2O_{(g)}$$

Scheme 19.1: Tungsten reduction.

This reaction is also not run stoichiometrically. Rather, an excess of hydrogen is used, the reaction is kept entirely oxygen free, and is run at elevated temperature. Excess hydrogen can be captured and re-used in the process until it is consumed. The tungsten product is collected as a powder form of the element.

The production of what are called ferrotungsten alloys does not necessarily take place after tungsten is reduced to the metal. Rather, patented processes involve the use of a tungsten ore, an iron-bearing ore, and some carbon-based reducing agent. As stated in one patent, what is required consists of: "A uniform mixture comprised of a finely-particulated tungsten-containing mineral, such as wolframite, scheelite, ferberite, and/or huebnerite; a supplemental quantity of a particulated iron-bearing material and a controlled amount of a carbonaceous reducing agent which is agglomerated into a plurality of pellets which are heated to an elevated temperature under a controlled vacuum for a period of time sufficient to effect a reduction of the tungstic oxide constituent to the metallic state and ... an alloying of the metallic tungsten with the iron constituent, producing substantially dense sintered ferrotungsten alloy pellets [6]." Thus, iron–tungsten alloys can be said to be formed in situ, without the two metals being combined only after being brought to their reduced states.

19.3 Uses

Very little elemental tungsten is used as a metal. Rather, more than half of reduced tungsten is used to produce tungsten carbide (WC), while much of the rest is utilized in various grades of steel. The tungsten filament that people think of as a common use, as part of a lightbulb, does indeed contain a certain amount of tungsten, but this and any other electrical uses are secondary in use to WC, at least in terms of volume.

19.3.1 Carbide parts

Tungsten carbide products, often abbreviated 'WC' because equal parts tungsten and carbon can be used, may be the end-user item with which people are most familiar. WC drill bits are high performance drills that can bore through metal, cement, and stone. The melting point of this material is 2,770 °C, which makes it very useful in any application that may result in a high temperature situation.

Beyond drills, WC can be made into other cutting tools, or used as a powder abrasive for sanding hard surfaces. As well, it has found use in the world's militaries as armor piercing rounds, most recently in anti-tank rounds. More recently, WC has been used in the jewelry industry, because it is highly resistant to scratches. WC can be molded into rings, although tungsten metal can as well. Thus, the idea of a tungsten wedding ring has arisen in the past few decades, and can now be found in many jewelry shops.

19.3.2 High-strength alloys

A wide variety of tungsten alloys exist, and all are used in some application that requires hardness and strength of the material. ASTM has an extensive set of standards for alloys containing tungsten [7].

What are sometimes called tungsten heavy alloys – those which also go by the names densimet, densalloy, or the trade name Mallory metal – contain nickel, copper, and iron as well as tungsten. Nickel is generally the second major components in such alloys. Such alloys tend to be used in automotive components where resistance to wear and corrosion are important.

19.4 Possible substitutes

When possible, molybdenum steels are substituted for tungsten steels, as long as the end product performance is as good as that of the item containing tungsten.

The tungsten filament light bulb, mentioned as one of the common uses people associate with tungsten has in the past twenty years had to compete with inexpensive light emitting diodes (LEDs) in a growing number of end-user applications.

19.5 Recycling

Tungsten alloy components of any user-end items are often recycled as part of the larger steel recycling industry. There are no programs for individual users to recycle tungsten-containing materials – none like the common household programs for glass, plastic, and paper – simply because there is much less tungsten or tungsten-containing materials used in houses and apartments.

Bibliography

[1] International Tungsten Industry Association. Website. (Accessed 16 April 2015, as: http://www.itia.info/).

[2] International Molybdenum Association. Website. (Accessed 4 December 2015, as: http://www.imoa.info/index.php).

[3] USGS Mineral Commodity Summaries. Downloaded as: http://minerals.usgs.gov/minerals/pubs/mcs/2015/mcs2015.pdf

[4] World Mineral Production, 2009–2013. British Geological Survey. (Downloadable from: http://www.bgs.ac.uk/mineralsUK/statistics/worldStatistics.html).

[5] Smith GR. United States Department of the Interior, Bureau of Mines Information Circular / 1994. 9388. Material Flow of Tungsten in the United States.

[6] Vacuum smelting process for producing ferrotungsten, US 4113479 A. United States Patent.

[7] ASTM. Website. (Accessed 2 December 2015, as: http://www.astm.org/Standards/E696.htm).

20 Tantalum and niobium

20.1 Introduction

Both tantalum and niobium (sometimes still called columbium amidst those in various professional organizations) are elemental metals with short histories, relatively small production quantities when compared to the largest scale industrial metals, and highly localized distribution throughout the planet, at least in terms of extractable material. First discovered at the turn of the nineteenth century, there was long term confusion lasting years as to whether the discovery represented one or two new elements.

Both elements were scientific curiosities for some time after their discovery, but in the past few decades they have become important enough in several applications, including in magnets and steels, that they are now tracked by several national and international organizations [1–4].

There have been a significant number of minerals identified that possess tantalum, niobium or both, although only a few of them are economically profitable to mine and from which to extract either of these two elements. Table 20.1 shows a list of tantalum- and niobium-containing minerals. Additionally, tantalum has been recovered as a secondary product from tin mining operations in various parts of the world.

Note that some of the mineral formulas shown in Table 20.1 are quite simple, while others are very complex. Despite specific formulas, a second level of complexity often accompanies the separation of a metal such as tantalum or niobium from an ore – the purity of the ore itself. In many cases, economically useless material, often silicates, must first be separated and the ore enhanced, or beneficiated.

Also of note, several of these minerals, especially coltan, are readily available in the eastern part of the Congo and surrounding countries in Africa. The ease of extraction there, coupled with existing wars in the region gave rise to the term 'resource curse'. This means that even though several countries are rich in coltan, extraction and export of the material goes to fund differing factions in the wars of the area, and destroys much of the traditional way of life of the local peoples [5].

Because both tantalum and niobium are used in several electronic and alloy applications, their global production is tracked annually by the USGS Mineral Commodity Summaries [1]. Figures 20.1 and 20.2 show respectively the production of tantalum and of niobium.

Despite the large number of minerals containing either tantalum and/or niobium, it is obvious from the two figures that economically exploitable deposits are not particularly widespread. Of both elements, the USGS states: "Domestic ... resources are of low grade, some are mineralogically complex, and most are not commercially recoverable [1]."

Table 20.1: Tantalum- and niobium-containing minerals.

Name	Formula	Location	Comments
Betafite	$(Ca, U)_2(Ti, Nb, Ta)_2O_6(OH)$	Canada	U, Th, and Nb can be extracted
Billwiseite	$Sb_5(Nb, Ta)_3WO_{18}$	Pakistan	Extremely rare
Coltan		Central Africa	A mixture of two minerals, columbite and tantalite
Columbite (aka. Niobite)	$(Fe, Mn)Nb_2O_6$	USA	
Euxenite	$(Y, Ca, Ce, U, Th)(Nb, Ta, Ti)_2O_6$	USA, Norway, Russia, Brazil, Sweden, Canada	
Microlite	$(Na, Ca)_2Ta_2O_6(O, OH, F)$	Sweden, USA	
Polycrase	$(Y, Ca, Ce, U, Th)(Ti, Nb, Ta)_2O_6$	Norway, USA	
Rynersonite	$Ca(Ta, Nb)_2O_6$	USA, Uganda	
Samarskite	$(Y, Fe, U, Th, Ca)_2(Nb, Ta)_2O_8$	Russia, Colorado	Has a Y and Yb form. Samarium found in an early deposit, named after the mineral
Simpsonite	$Al_4(Ta, Nb)_3O_{13}(OH)$	Brazil, Australia	
Tantalite	$(Fe, Mn)Ta_2O_6$	USA, Australia, Brazil, Canada, Nigeria, Rwanda, Zimbabwe	
Tantite	Ta_2O_5	Russia	Rare mineral
Tapiolite	$(Fe, Mn)(Nb, Ta)_2O_6$	Finland	Has an Mn-rich and an Fe-rich form
Wodginite	$Mn(Sn, Ta)Ta_2O_8$	Australia, Canada, USA, Brazil, Finland, Kazakhstan	Can also include Nb
Yttrocolumbite	$(Y, U, Fe)(Nb, Ta)O_4$	Mozambique	
Zimbabweite	$(Na, K)_2PbAS_4(Nb, Ta, Ti)_4O_{18}$	Zimbabwe	First discovered in 1986

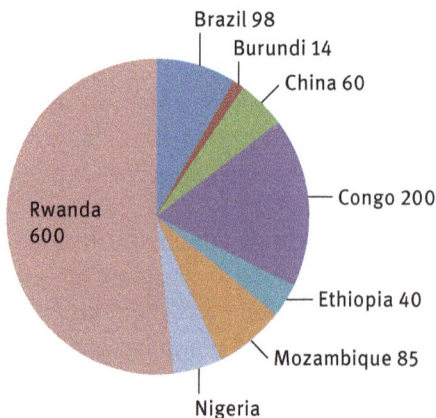

Fig. 20.1: Tantalum production, in metric tons.

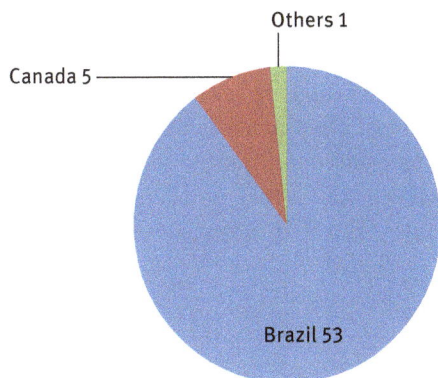

Others 1

Canada 5

Brazil 53

Fig. 20.2: Niobium production, in metric tons.

20.2 Refining and isolation

As seen, there are several minerals that contain small amounts of tantalum or niobium, often along with some other metal or metals, but these are usually mined with another element as the primary product. The single material with the highest concentration of both of these elements is coltan, listed in Table 20.1, a mixture of the two minerals columbite and tantalite.

As with many of the metals discussed in the previous chapters, tantalum- and niobium-bearing minerals often must be concentrated before the extraction and reduction chemistry can be undertaken. These processes can be complex, but ultimately the reaction chemistry in which each metal is formed into a complex can be written in a somewhat simplified form, as shown in Scheme 20.1.

$$Ta_2O_5 + Nb_2O_5 + 24\,HF \rightarrow 2\,H_2[TaF_7] + 2\,H_2[NbOF_5] + 8\,H_2O$$

Scheme 20.1: Complex formation for tantalum and niobium.

The two complexes can then be separated using an organic solvent such as cyclohexanone for the extraction. This takes advantage of different solubilities of the different complexes, as shown in Scheme 20.2.

$$2\,KF + H_2[NbOF_5] \rightarrow 2\,HF + K_2[NbOF_5]_{(s)}$$

or

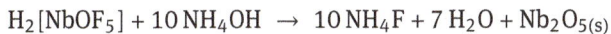

$$H_2[NbOF_5] + 10\,NH_4OH \rightarrow 10\,NH_4F + 7\,H_2O + Nb_2O_{5(s)}$$

and

$$H_2[TaF_7] + 7\,NH_3 + 5\,H_2O \rightarrow Ta(OH)_5 + 7\,NH_4F$$

Scheme 20.2: Tantalum and niobium separation.

From this point there is more than one method by which tantalum can be reduced to the metal, but sodium is often the reducing agent, and the reaction is run at elevated temperatures (around 800 °C). Niobium can also be reduced to the metal after separation from tantalum, but it is sometimes advantageous to produce ferroniobium alloys directly from the niobium oxide. This is done using aluminum as the reducing agent, and is thus a form of Goldschmidt reaction, in which oxygen is transferred from one metal to another. The reaction chemistry for this is shown in Scheme 20.3.

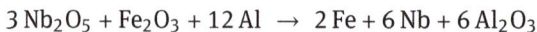

$$3\,Nb_2O_5 + Fe_2O_3 + 12\,Al \rightarrow 2\,Fe + 6\,Nb + 6\,Al_2O_3$$

Scheme 20.3: Ferroniobium production.

Once again, this reaction must be run at elevated temperatures to ensure formation of the alloy. The amount of niobium in the final alloy is determined by the ratio of niobium oxide to iron oxide in the original reaction mixture. Often, alloys that are as high as 60 % niobium are produced.

20.3 Uses

20.3.1 Major uses

Tantalum and niobium each have several specialized uses, listed in in Table 20.2. Some uses that are not listed in the table are also discussed because they are smaller volume but continue to serve specific needs.

Table 20.2: Tantalum and niobium uses [1]

Metal	Use	Comment
Tantalum	Capacitors	Greater than 60 % of Ta use, for cell phones, computers, automotive electronics
	Alloys	
	Metal powder	
	Ta_2O_5	Finds use in specialty glass lenses
	TaC_x	In cutting tools
Niobium	Ferroniobium	Major use, in steels, ca. 60 %
	Superalloys	Name indicates high hardness and wear resistance

20.3.2 Wiring and magnets

The best-known use for niobium-containing wire is familiar to those in scientific laboratories, because it is used in nuclear magnetic resonance (NMR) spectrometer magnets. A high field, superconducting magnet (300 MHz and above) can utilize several miles of wire. There are few companies capable of producing such a wire to the standards needed for the production of a high field, superconducting electromagnet, and all claim their fabrication technique is proprietary. However, the formula $(NbTaTi)_3Sn$ is known to have been used in producing such wire. One of the current NMR manufacturers, Jeol, states at its website that a niobium–titanium wire has been used in their magnets [6]. Additionally, niobium alloyed with tin or with aluminum has been found to form superconducting wire, which again can be used to produce superconducting magnets.

20.3.3 Niobium coinage

The Austrian Mint has for slightly more than a decade produced a series of collector coins made from exotic metals, including some made from what is called a bimetallic composition of silver and niobium. This term indicates that the two metals are not alloyed. Rather, one forms a central ring, and the other an outer ring. The coins are not designed to be used in daily commerce, but rather are marketed directly to collectors. Thus, they bear some face value, but it is always much lower than the market value of the two metals that make up the coin [7].

Additionally, Latvia has produced several of these bimetallic coins which utilize niobium as one of the two metals. In 2014, Luxembourg also began to produce a series of such coins.

Also, starting in 2011, the Royal Canadian Mint in Ottawa began producing coins containing niobium for the collector market. The RCM is one of the larger mints in the world, and thus more niobium may be used in collector coinage in coming years, if these coins prove popular with collectors. Much like those from the Austrian Mint, these Canadian pieces have some face value that is significantly lower than the value of the metal or metals from which they are made [8].

20.4 Possible substitutes

Some metal elements can substitute for both tantalum and niobium in their alloys, but the USGS notes that this may result in a material that does not perform to the same standards [1]. Beyond this, when these two metals are used in NMR magnets, there really is no substitute alloy that has yet been found that produces the same field.

20.5 Recycling

The recycling of ferroniobium is a part of the larger steel and iron recycling industry. The decommissioning of an NMR usually means the metal in it will be re-used or recycled if needed, which includes the wiring. The two different Mint programs for collector coins that incorporate niobium are young enough that there has never been any call to re-melt and recycle such items (there have in the past been gold and silver melts in terms of coins, when the price of either metal spikes on the world's markets).

Bibliography

[1] USGS Mineral Commodity Summaries. Downloaded as: http://minerals.usgs.gov/minerals/pubs/mcs/2015/mcs2015.pdf
[2] Tantalum Niobium International Study Centre. Website. (Accessed 16 April 2015, as: http://www.tanb.org/).
[3] Minor Metals Trade Association. Website. (Accessed 6 December 2015, as: http://www.mmta.co.uk/home).
[4] Global Advanced Metals. Website. (Accessed 28 April 2015, as: http://www.globaladvancedmetals.com/news/announcements/2011/january/world%E2%80%99s-largest-tantalum-producer-resumes-operations.aspx).
[5] Institute for Environmental Security. Coltan Mining in the Democratic Republic of Congo. Website. (Accessed 6 December 015, as: http://www.envirosecurity.org/actionguide/view.php?r=207&m=publications).
[6] Jeol. Website. (Accessed 14 December 2015, as: http://www.jeolusa.com/RESOURCES/AnalyticalInstruments/NMRMagnetDestruction/tabid/390/Default.aspx).
[7] Austrian State Mint. Website. (Accessed 16 December 2015 as: https://www.muenzeoesterreich.at/eng).
[8] Royal Canadian Mint. Website. (Accessed 16 December 2015, as: http://www.mint.ca/store/template/home.jsp).

21 Sodium

21.1 Introduction

Few metals have as short a time span between their discovery and a major industrial use as sodium. First discovered in 1807, sodium was difficult to isolate until what is called the Deville process was discovered and developed, in which sodium carbonate is reacted with carbon as a reducing agent, to produce sodium metal and carbon monoxide. This is shown in Scheme 21.1.

$$Na_2CO_3 + 2\,C \;\rightarrow\; 3\,CO_{(g)} + 2\,Na$$

Scheme 21.1: Deville process for sodium reduction.

The Deville process, while effective and useful for years, has been superseded by the Downs cell operations currently in use. Downs cell apparatuses were first introduced in the 1920s.

In several of the earlier chapters tables of minerals were presented to illustrate the many sources of a particular element. There are several minerals that contain sodium, but all sodium metal is now produced from common salt, NaCl, which is considered a virtually inexhaustible resource. Very few countries do not have access to salt, either from evaporation of ocean waters, brine lakes and inland seas, or what is still called halite, meaning salt from mines. Historically, halite has been mined in eastern Africa, southern Germany, western Canada, and near Cleveland, Ohio, and Detroit, Michigan in the USA. Figure 21.1 shows the world production of salt, from which sodium metal is reduced [1].

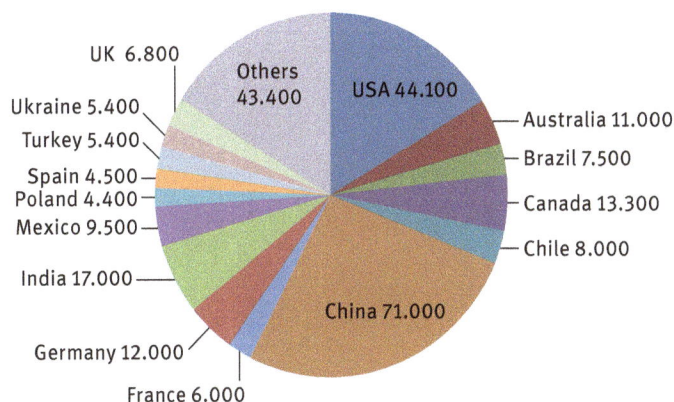

UK 6.800
Others 43.400
USA 44.100
Ukraine 5.400
Turkey 5.400
Australia 11.000
Spain 4.500
Brazil 7.500
Poland 4.400
Mexico 9.500
Canada 13.300
India 17.000
Chile 8.000
Germany 12.000
China 71.000
France 6.000

Fig. 21.1: World salt production, in thousands of metric tons.

The amounts of sodium chloride used annually (common table salt) are much greater than that of sodium metal, and consequently, organizations such as the United States Geological Survey track the production and uses of salt, and not that of sodium metal or chlorine gas, two major products derived from salt [1].

21.2 Refining and isolation

The two largest chemical reactions involving sodium, both starting with sodium chloride, are the reduction of it to sodium metal, and its use in the production of sodium hydroxide. Numerous companies worldwide produce sodium metal, with several of them advertising the purity of their product [2, 3].

21.2.1 Downs cell

The Downs cell operation is an electrolytic oxidation–reduction reaction that produces sodium metal and chlorine gas from molten sodium chloride. Because sodium chloride must be in a liquid state in order for the reaction to occur, it must be run in a refractory-lined container (essentially a brick container), because the melting point of sodium chloride is 801 °C. Routinely, the process requires a carbon anode for the oxidation from chloride to elemental chlorine, and an iron cathode for the reduction of

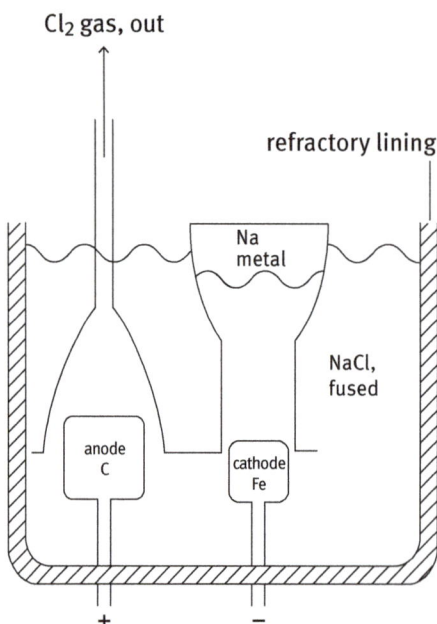

Fig. 21.2: Downs cell.

sodium from its cationic state to the reduced metal. Also, calcium chloride is present in the melt, because the mixture of the two salts has a lower melting point – about 700 °C – than that of pure sodium chloride. This translates to a cost savings in production because less energy is required. Figure 21.2 shows a schematic diagram of a Downs cell.

The reaction chemistry for a Downs cell can be written very simply, as shown in Scheme 21.2. The calcium chloride in the mixture does not enter into the reaction.

$$2\,NaCl_{(l)} \rightarrow 2\,Na_{(l)} + Cl_{2(g)}$$

Scheme 21.2: Downs cell reaction.

21.2.2 Chlor-alkali process

The chlor-alkali process, one of the largest commodity chemical processes in the world, routinely utilizes sodium chloride to make sodium hydroxide – often called industrial caustic – as well as chlorine gas and hydrogen gas. There are three main variations of the process, the diaphragm cell, the membrane cell, and the mercury cell, and none isolate elemental sodium. The reaction chemistry is the same for all three versions of the process, and is shown in Scheme 21.3.

$$2\,NaCl_{(aq)} + 2\,H_2O_{(l)} \rightarrow 2\,NaOH_{(aq)} + Cl_{2(g)} + 2\,H_{2(g)}$$

Scheme 21.3: Chlor-alkali chemistry.

The sodium hydroxide that is isolated as an aqueous solution via the mercury cell can be as high as 50 % industrial caustic, and can be further concentrated from that point. This process actually amalgamates the sodium into the mercury, which acts as the cathode, but the reduced sodium is not isolated. It is re-oxidized in a chamber separate from the chlorine, and recombined with water to form NaOH. Since our focus is on metals and alloys, this process will not be treated in further detail here.

21.3 Uses

Elemental sodium is usually converted into several compounds that are economically valuable, although there are some uses for elemental sodium metal. The production and uses of these compounds and the metal are as follows:

21.3.1 Sodium borohydride

Production of sodium borohydride is through sodium hydride, which is made in a direct elemental combination, as shown in Scheme 21.4 [4–6].

$$2\,Na_{(l)} + H_{2(g)} \rightarrow 2\,NaH$$

Scheme 21.4: Sodium hydride production.

Sodium borohydride must be prepared by treating sodium hydride with trimethyl borate at an elevated temperature, usually 250–270 °C, as shown in Scheme 21.5.

$$4\,NaH + B(OCH_3)_3 \rightarrow NaBH_4 + 3\,NaOCH_3$$

Scheme 21.5: Sodium borohydride production.

A major use is bleaching various types of wood pulp. Indeed, one firm, Montgomery Chemicals, states at its website: "Montgomery Chemicals is a leader in Borohydride reducing agents for use in the pulp and paper, pharmaceutical, organic chemical purification, metal recovery, textile and fuel cell industries [6]."

21.3.2 Sodium azide, NaN_3

Sodium azide is produced through a two-step process, as shown in Scheme 21.6 [7]. Its large-scale production has a history that is approximately two decades old.

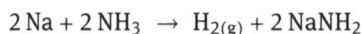

$$2\,Na + 2\,NH_3 \rightarrow H_{2(g)} + 2\,NaNH_2$$

followed by

$$N_2O_{(g)} + 2\,NaNH_2 \rightarrow NaN_3 + NH_3 + NaOH$$

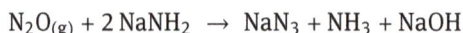

Scheme 21.6: Production of sodium azide.

The major use of the material is in automotive airbags, although it also finds use in inflatable slide chutes on passenger airplanes, and in the production of other azides. Although the amount used in a single airbag is quite small (under 100 grams), the number of airbags produced annually makes the production of sodium azide a multi-ton per year process [7].

The production of sodium amide – the first of the two reactions in Scheme 21.6 – has also been used in the production of synthetic indigo dye.

21.3.3 Triphenylphosphine, $P(C_6H_5)_3$

Triphenylphosphine, often abbreviated to PPh_3 or TPP, is made on a large scale according to the reaction shown in Scheme 21.7.

$$PCl_3 + 6\,Na + 3\,C_6H_5Cl \rightarrow P(C_6H_5)_3 + 6\,NaCl$$

Scheme 21.7: Triphenylphosphine production.

Triphenylphoshine then finds uses in numerous organic and organometallic reactions. There are several producers of it. One, BASF, states at its website: "TPP is a highly efficient product that serves successfully in many applications, for example: in vitamin synthesis and production of pharmaceutical active ingredients, crop protection products and coatings, as a co-catalyst in isobutanol and n-butanol production, as an initiator of several polymerization reactions, as an oxidation and UV stabilizer in plastics [8]."

21.3.4 Reactor moderator material

Sodium metal can be used in nuclear reactors as a coolant, generally called a heat transfer fluid, and has some advantages over water in this role. However, more care must be taken when sodium metal is used in this role because of its reactivity with water. Additionally, maintenance of reactor cores is somewhat more difficult when liquid sodium is used, because a person cannot see through it, as one can with water.

21.3.5 Alloying agent

Sodium–potassium alloys, often called NaK alloys, find use as drying agents in inert atmosphere boxes, because the alloy is a liquid at room temperature. There can be several different sodium–potassium alloys, but the eutectic point for the binary mixture occurs at 77 % potassium. Additionally, nuclear reactors can utilize NaK alloys at lower temperatures, since most NaK alloys are a liquid at ambient temperature.

21.3.6 Sodium vapor lamps

What are called sodium vapor lamps actually function by exciting a sodium–mercury amalgam in the gas phase. Through manipulation of pressures and voltages, a variety of colors can be obtained. The popular yellow street light is a form of low pressure sodium vapor lamp.

21.4 Recycling

Sodium is an extremely reactive metal, and industries that use large amounts of it have stringent safety programs for their employees, so that proper handling and monitoring is ensured at all times. In part because of this reactivity, but also because so much sodium is used to produce other chemical intermediates or end use materials, there are no recycling programs for it.

Bibliography

[1] USGS Mineral Commodity Summaries. Downloaded as: http://minerals.usgs.gov/minerals/pubs/mcs/2015/mcs2015.pdf
[2] Lantai Industry. Website. (Accessed 23 June 2015, as: http://www.lantaicn.com/doce/view.asp?auto_id=138).
[3] DuPont. Website. (Accessed 23 June 2015, http://www2.dupont.com/Reactive_Metals/en_US/).
[4] US Patent. Preparation of sodium hydride from sodium amalgam and hydrogen, US 2829950 A.
[5] TNJ Chemicals. Website. (Accessed 23 June 2015, as: http://tnjchem.com/Article1.Asp?Pid=194&id=1696).
[6] Montgomery Chemicals. Website. (Accessed 23 June 2015, as: http://www.montgomerychemicals.com/?gclid=ClvvwrXHpsYCFQuMaQodU9YFpQ).
[7] Grainger. Website. (Accessed 23 June 2015, as: https://www.grainger.com/search?searchQuery=Sodium+Azide).
[8] BASF. Website. (Accessed 23 June 2015, as: http://www.intermediates.basf.com/chemicals/triphenylphosphine/index).

22 Lithium

22.1 Introduction

Lithium, like many other elemental metals, was discovered in the nineteenth century. In its short history, it has undergone several changes in demand, because the end products made with it have waxed and waned in demand. For the most recent example, the now-popular lithium batteries were not commercially produced on a large sale prior to the 1980s.

22.2 Refining and isolation

Several different minerals contain some amount of lithium, but spodumene ($LiAlSi_2O_6$) is the major mineral from which lithium metal has been produced profitably. The demand for lithium batteries has also fueled a search for lithium sources of all kinds, and thus in some processes, lithium from brine solutions has overtaken spodumene as source material for lithium products in the past few years.

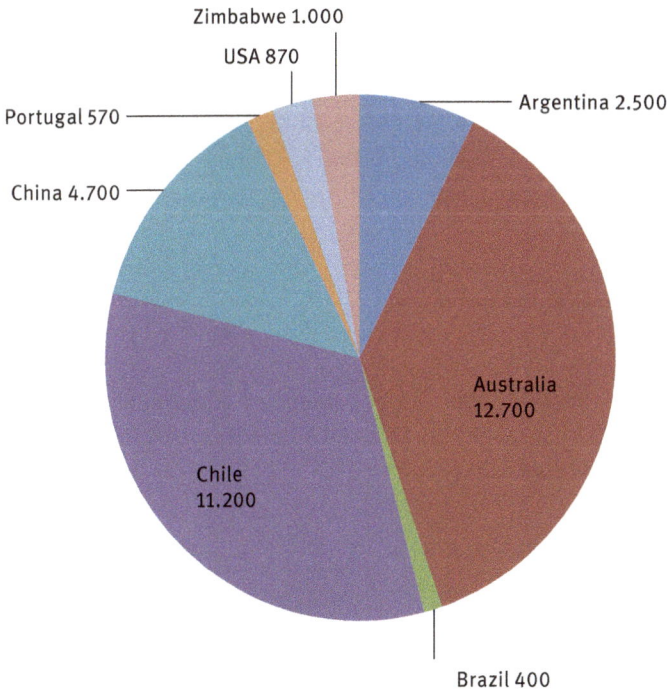

Fig. 22.1: Lithium production, in metric tons.

Lithium occurs in localized areas, although some of these areas, like the Salar de Uyuni in Bolivia, are quite large. Lithium production is tracked by the USGS Mineral Commodity Summaries, as well as the British Geological Survey, and is shown in Figure 22.1 [1, 2].

Note that Chile and Argentina are both represented in Figure 22.1. These nations draw from the same general source for lithium, the Atacama High Desert and what are called the salars – the seas – that occur seasonally within it [2]. Bolivia is not represented, although it probably will be in the near future, as it has large reserves of lithium that have not yet been brought to the market. In the United States, lithium has in the past been extracted from deposits at Searles Lake, Nevada. In the future, this may change and include sources from the state of Wyoming, where new deposits have reportedly been found. The large amounts produced in China and Australia are generally from spodumene ores. Finally, one nation that is conspicuously absent from Figure 22.1 is Afghanistan. Reports from the time of the Soviet occupation indicate there may be large deposits of lithium-containing materials there, but the decades of war have prevented any development and extraction [3].

The production of lithium metal from spodumene is complex, and involves the usual pre-chemical steps of concentration of the ore and separation from rock and other unwanted material. However, lithium reduction from the salts in brine solutions is somewhat more direct. Scheme 22.1 shows the simplified reaction chemistry starting with different salts.

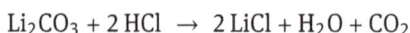

$$Li_2CO_3 + 2\,HCl \;\rightarrow\; 2\,LiCl + H_2O + CO_2$$

or

$$LiOH + HCl \;\rightarrow\; LiCl + H_2O$$

followed by

$$2\,LiCl_{(l)} \;\rightarrow\; Li_{(l)} + Cl_{2(g)}$$

Scheme 22.1: Reduction from lithium salts to lithium metal.

This final reaction runs at over 600 °C, but can be lowered by addition of KCl to approximately 450 °C. Usually a mixture of 55 % LiCl and 45 % KCl is used as that is the eutectic point for LiCl–KCl mixtures.

22.3 Uses

Very little reduced lithium metal is refined, then used as the metal, at least when compared to lithium used in compounds. This is because it is very reactive, and unlike elemental sodium, does not have any practical use as a metal. The USGS Mineral Commodity Summaries summarizes lithium use as follows: "Ceramics and glass, 35 %;

batteries, 31 %; lubricating greases, 8 %; continuous casting mold flux powders, 6 %; air treatment, 5 %; polymer production, 5 %; primary aluminum production, 1 %; and other uses, 9 % [1]." These uses are essentially those of lithium compounds, meaning lithium in the +1 oxidation state.

However, lithium alloys do see use on an industrial scale because they reduce weight and increase hardness. We will briefly discuss a few lithium compounds.

22.3.1 Lightweight alloys

While lithium is part of several different alloys, these are usually considered to be aluminum alloys, because there is more aluminum in them than any other metal. Lithium's density is $0.534\,\mathrm{g/cm^3}$, which makes it the least dense metal, and thus a useful one for applications where weight matters, such as the aerospace industry. But aluminum's density is $2.70\,\mathrm{g/cm^3}$, which is still a rather light metal (for a point of comparison, the density of iron is $7.874\,\mathrm{g/cm^3}$). Since aluminum is produced on such a large scale, and is economically profitable to make in such large amounts, most applications where less weight is critical are treated using aluminum alloys.

22.3.2 Batteries

Lithium batteries now encompass a wide variety of different materials, few of which include reduced lithium metal, because of its reactivity. For example, two organic solvents that have been used in lithium primary cells (those that do not recharge) as well as secondary cells (those that are rechargeable), propylene carbonate and dimethoxyethane, have been found to be compatible with lithium even if there is accidental over-reduction to lithium metal. In very common terms, these solvents prevent a battery from accidentally catching fire if lithium metal were to form.

What has become the older use of lithium batteries – small devices that require small, light batteries – has been overtaken in the past five years in terms of demand for lithium by automobile manufacturers and the batteries used in electric cars. The reason lithium batteries are so desirable for electric automobiles is a matter of what is called charge density. Essentially, since lithium has such low density, batteries that produce the same power as those made with other materials do not weigh as much, and thus electric cars are able go farther because they weigh less using such batteries.

There is more than one type of what can be called a lithium battery. For automobiles, the most common is the lithium-ion battery [4]. Its reaction chemistry for charge and discharge can be represented in a somewhat simplified manner as shown in Scheme 22.2.

$$C_6 + LiCoO_2 \rightarrow Li_xC_6 + Li_{1-x}CoO_2$$

followed by

$$Li_xC_6 + Li_{1-x}CoO_2 \rightarrow Li_{x-y}C_6 + Li_{1-x+y}CoO_2$$

Scheme 22.2: Lithium-ion battery cycle.

The x represents loss of a non-stoichiometric amount of lithium from its starting point, a lithium cobalt oxide, into graphite. The y represents the return of some non-stoichiometric amount of lithium back to the lithium cobalt oxide. As the cycle runs, x is always less than 1, and y is always less than x.

Importantly for this discussion, lithium metal in the reduced state does not participate in these reactions. Perhaps obviously however, lithium is required for the synthesis of the lithium cobalt oxide.

Two other lithium batteries, ones that do utilize lithium metal, are shown in Scheme 22.3. In each case, the solvent is not water. In the first, thionyl chloride and sulfur dioxide function as the solvent. In the second, sulfur dioxide acts as the solvent. In both cases, the battery is under pressure to keep the solvent a liquid.

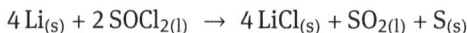

$$4\,Li_{(s)} + 2\,SOCl_{2(l)} \rightarrow 4\,LiCl_{(s)} + SO_{2(l)} + S_{(s)}$$

as well as

$$2\,Li_{(s)} + 2\,SO_{2(l)} \rightarrow Li_2S_2O_{4(s)}$$

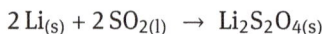

Scheme 22.3: Lithium batteries utilizing lithium metal.

Both of these are primary batteries, and thus are not of use as automotive batteries. They have however found use in military applications, and in spacecraft.

22.4 Possible substitutes

Aluminum alloys containing lithium now compete with other lightweight alloys, including those that use small amounts of boron, or more recently, materials made from nonmetal, polymer composites.

22.5 Recycling

Lithium battery recycling is part of the larger idea of recycling all batteries. In many locations, this remains a voluntary effort. In the United States, there is not yet any national law of regulation regarding this type of recycling.

Automobile manufacturers who are now marketing electric vehicles are planning on re-using lithium batteries from their cars when such cars are traded in or in some other way decommissioned. The use may not be in a second automobile, rather, it may be in some large-scale power transfer system. It is estimated that recycling the lithium battery from an electric vehicle is significantly less expensive than producing a new one from raw materials.

Bibliography

[1] USGS Mineral Commodity Summaries. Downloaded as: http://minerals.usgs.gov/minerals/pubs/mcs/2015/mcs2015.pdf
[2] British Geological Survey, World Mineral Production. Downloadable at: http://www.bgs.ac.uk/mineralsUK/statistics/worldStatistics.html
[3] USGS, Summaries and Data Packages of Important Areas for Mineral Investment and Production Opportunities in Afghanistan, downloadable at: http://pubs.usgs.gov/fs/2011/3108/fs2011-3108.pdf
[4] Western Lithium WLC. Website. (Accessed 17 December 2015, as: http://www.westernlithium.com/western-lithium-announces-post-merger-integration-and-organization-update-appoints-new-executive-team-and-new-director/).

Index

A
Aich's alloy 10
alumina 73
aluminum 71
amalgam 104
American Wire Gauges 8
americium 97
americium dioxide 98
aqua regia 57
Atacama High Desert 146

B
basic oxygen furnace 48
bauxite 72
Bayer process 73
Becquerel, Henri 89
Betterton–Kroll process 119, 120
Betts electrolytic process 120
Betts process 119
blast furnace 46
brass 5, 10
brine lakes 139
bronze 5, 10
Bronze Age 1
Bushveld Igneous Complex 55

C
carat 33
carbon dioxide 51
Carbon in Pulp process 32
carbon monoxide 51
Carbonyl process 66
catalytic converter 61
Cativa process 59
Cazo process 38
centrifugal concentration 112
ceria 113
chlor-alkali process 103, 141
chlorine 103
cinnabar 101, 102
cis-platin 62
cobalt 65
coin silver 9
coltan 133, 135
columbium 133

copper 5
Curie, Marie 89
Cyanide process 32

D
Damascus steel 49
Davy, Sir Humphry 83
depleted uranium rounds 93
Deville process 139
dolomite 83
Doré bars 33
Dow process 84
Downs cell 139, 140

E
East Asian Tin Belt 13
electric arc furnace 48
electromagnetic separation 111
electrowinning 8, 20
electrum 29
Epsom salts 85
Euro coins 70

F
ferrotungsten alloys 130
Feuchtwanger 22
Float process 16
froth flotation 20

G
galena 119
galvanizing 21
gangue 20
gold 29
Goldschmidt reaction 136

H
Hall–Heroult process 71
heavy rare earth element 115
hematite 45
Hunter process 80
hydrogen 103

I
ilmenite 77
Industrial Revolution 1
iridium 59

iron 43
Iron Age 1

K
Kroll process 79, 85
Krugerrand 34

L
lanthanides 92, 107
lead 117
– acid batteries 121
– poisoning 117
– sheet 121
light emitting diodes 41
lightweight alloys 147
limestone 15
lithium 145
– batteries 147

M
MacArthur–Forrest process 32
mag wheels 87
magnesium 83
magnetic separation 129
magnetite 45
mercury 101
– oxide cells 104
– poisoning 101
Merensky Reef 55
Miller process 32
molybdenum steels 131
Mond process 66
Muntz metal 10

N
nickel 65
nickel silver 21
niobium 133
Nordic gold 9
nuclear magnetic resonance 137

O
osmium 59

P
palladium 60
Parkes process 39, 119, 120
Patio process 38
periodic table 2

pewter 11, 25
pewterware 26
Pilkington process 16
pitchblende 89
placer copper 6
platinum group metals 55
polyethylene oxide 41
polyvinyl chloride 16
porphyry 6
Potosi mines 14

Q
quicksilver 101

R
radioactivity 89
rare earth elements 92
rare earth oxide (REO) equivalents 110
recycling 4
resource curse 133
rhodium 59
ruthenium 58, 81
Rutherford, Ernest 89
rutile 77

S
sacrificial anode 23
salars 146
samarium–cobalt magnets 114
sand casting 121
shape memory alloys 68
silver 37
– bullion coins 40
silver oxide battery 23, 41
slag 46, 47, 51
smelting 7
smoke detector 98
sodium 139
– azide 142
– borohydride 142
– hydroxide 103
– potassium alloys 143
– vapor lamps 143
solders 15, 22, 122
solid oxide membrane 85
specimen silver 69
spodumene 145
steel 43
super magnets 113
superalloy 67, 136

T
taconite 45
tantalum 133
thermite reaction 74
thermometer 104
thoria 94
thorium 89, 107
tin 13
titania 79
titanium 77
– sponge 77, 80
– white 81
Toledo steel 50
Tombac 10
triphenylphosphine 143
tungsten 125
– carbide 130
– heavy alloys 131
– steels 131

U
uranium 89
uranium-235 93

V
vermillion 104

W
Wetterhahn, Karen 101
Wharton, Joseph 68
white gold 35
Wohlwill process 33
Wootz steel 49

Y
yellow cake 90

Z
zinc 19
zinc white 22

www.ingramcontent.com/pod-product-compliance
Lightning Source LLC
Chambersburg PA
CBHW081108220326
41598CB00038B/7271